THIS IS YOUR BRAIN

TEACHING
ABOUT
NEUROSCIENCE
AND
ADDICTION
RESEARCH

THIS IS YOUR BRAIN

TEACHING ABOUT NEUROSCIENCE AND ADDICTION RESEARCH

TERRA NOVA
LEARNING SYSTEMS

National Science Teachers Association

Arlington, Virginia

National Science Teachers Association

Claire Reinburg, Director
Jennifer Horak, Managing Editor
Andrew Cooke, Senior Editor
Wendy Rubin, Associate Editor
Agnes Bannigan, Associate Editor
Amy America, Book Acquisitions Coordinator

Art and Design
Will Thomas Jr., Director
Joe Butera, Senior Graphic Designer, cover and interior design
Cover photograph by Joan Cantó

Printing and Production
Catherine Lorrain, Director
Nguyet Tran, Assistant Production Manager

National Science Teachers Association
Francis Q. Eberle, PhD, Executive Director
David Beacom, Publisher
1840 Wilson Blvd., Arlington, VA 22201
www.nsta.org/store
For customer service inquiries, please call 800-277-5300.

Library of Congress Cataloging-in-Publication Data is available from the Library of Congress.

CONTENTS

PREFACE

The original idea for *This Is Your Brain: Teaching About Neuroscience and Addiction Research* was developed by two physician-scientists. The curriculum was proposed to the National Institutes of Health and underwent review and critique by scientific experts. The curriculum was seen as a priority, and our multidisciplinary team was encouraged (and funded) to develop it. The curriculum was overseen and approved by a scientist who also serves as a program officer at the National Institute on Drug Abuse. The creation of the educational materials was informed by up-to-date scientific information, and scientists and science teachers methodically evaluated the components of the curriculum. Wherever possible, we integrated and fulfilled national standards for middle school education (e.g., National Science Education Standards, National Council of Teachers of Mathematics, National Council of Teachers of English, Consortium of National Arts Education Associations, and others). Once these steps were completed, the curriculum was reviewed and tested by classroom teachers, by parents, and by students. The curriculum was then revised on the basis of all of this feedback, and the ultimate result is this book, the Mouse Maze, and all of the curricular companion materials included in *This Is Your Brain*.

Our aim in developing this work is to inform young people about the value of science in our world. Our intent is that this curriculum will help support independent and intelligent thinking about some of the very hardest issues in science and society—what questions to pursue, how scientific work is performed, and how science is conducted ethically. Our hope is that this work will inspire many young people to become the scientists who will answer important questions that will lead to healthier lives and a healthier world in the future.

No single individual authored this book, and it does not represent the perspective, approach, or opinion of a single individual. This project was undertaken as a collaborative effort that was based in a small company, Terra Nova Learning Systems (TNLS). TNLS is owned by one of the physician-scientists who conceptualized and served as a principal investigator for this project. At the time in which *This Is Your Brain* was being developed, TNLS was located in Milwaukee, Wisconsin; it subsequently has moved to Palo Alto, California.

This work is dedicated to Helen, Willa, and Tommy Roberts, and to the memory of their father, Brian B. Roberts, M.D., the other physician-scientist who led the Terra Nova Learning Systems team in creating this curriculum.

SECTION ONE
HOW TO UNDERSTAND AND USE *THIS IS YOUR BRAIN*

10-LESSON UNIT DESCRIPTION

Welcome to *This Is Your Brain: Teaching About Neuroscience and Addiction Research*, an innovative curriculum that creatively explores contemporary real-world issues in science and research. Funded by the National Institute on Drug Abuse (NIDA) and developed in consultation with nationally recognized experts, this stimulating unit presents 10 comprehensive, ready-to-use lessons for the middle school classroom. Each 45-minute lesson incorporates National Science Education Standards and purposeful objectives to sharpen important decision-making and higher-order thinking skills.

Following scientific method principles, the lessons in this book guide students through concepts such as brain structure and function, the neurobiology of drug addiction, the role of biomedical research in understanding addiction and prevention, and the ethics of animal inclusion in biomedical research. Filled with skill-based, multicurriculum activities, this unit is ideal for use by middle school science, health, physical education, and family consumer science teachers as well as guidance counselors, social workers, psychologists, and nurses.

INTRODUCTION

Insights from neuroscience are changing how we live each day, and will create a better future for our world. Neuroscience helps us to understand the underlying processes of life and healthy development, and it helps us to know why we may become vulnerable to disease. It also helps to identify ways of preventing, diagnosing, and treating serious illnesses and conditions.

One example is the Nobel Prize–winning neuroscience work from more than 60 years ago that has led to the near-elimination of polio in North America. Polio most commonly affects children and is caused by a virus that kills cells in the spinal cord and brain. When these cells die, the body loses the functions served by these cells, such as motor movement or breathing. Depending on what cells are affected, a person may experience disability (e.g., paralysis of an arm or a leg) or death (e.g., loss of respiratory function). Neurological research led to the identification of the virus that causes polio and gave insights that led to a vaccine effective in neutralizing the virus before it attacked the spinal cord and brain.

More recently, neuroscience has led to a far greater understanding of brain development; of the circuitry and plasticity of the brain; and of the functions of the brain, such as imagination and memory. Neuroscience has also led to greater knowledge of health problems including addic-tion, autism, bipolar disorder, dementia, depression, infectious diseases such as HIV and "Mad Cow disease," multiple sclerosis, Parkinson's disease, schizophrenia, and stroke. Neuroscience has facilitated the development of new treatments (and the calculation of when treatment interventions will be optimally successful). And amazing new neuroscience research has led to previously unimaginable approaches to rehabilitation. For example, Stanford researchers have developed a way in which people can draw and write on a computer screen just through *thinking* about drawing and writing, without using their hands. These studies are now being tried with people paralyzed due to catastrophic spinal cord injuries—with incredible results.

In *This Is Your Brain*, we have focused on neuroscience and the specific area of drug abuse research. Drug use and abuse has become a substantial public health burden in our country in terms of personal suffering and societal impact, with the greatest impact on today's young people. A recent study, the National Survey on Drug Use and Health (NSDUH), found that over 19 million Americans aged 12 or older were current (past month) illicit drug users, with marijuana use being most prevalent. Moreover, the study reported 3.8% of youths ages 12 or 13 identified themselves as current illicit drug users, 10.9% for youths ages 14 or 15, and 17.3% for youths ages 16

4

or 17, with use peaking at 21.7% among 18- to 20-year-olds. In another study, the 2009 Monitoring the Future survey, 6.6% of 12th graders admitted to using methamphetamines in the prior year. The CDC's most updated figures suggest that 8.7% of people age 12 and older used illicit drugs in the prior month. The abuse of legal prescription medications has also increased greatly, with recent estimates of 2.8% of people age 12 and older misusing psychotherapeutic agents for nontherapeutic purposes in the prior month. These statistics indicate an urgent need for preventative measures, specifically drug abuse education, to deter drug use by our country's youth.

The economic costs associated with drug use, abuse, and addiction are staggering. The devastating effects influence all aspects of personal and social life, including healthcare, crime, social welfare, lost productivity, infant morbidity and mortality, and premature death. With access to addictive substances at an all-time high, students need a foundational understanding of the underlying causes of and reinforcements to drug abuse in order to make responsible, informed, ethical decisions regarding drug use prevention, research, and treatment.

It is critically important to focus young people's attention on efforts to deter illicit drug use because their risk of exposure to drugs increases substantially as they pass from elementary school to middle and high school. Children in early and middle adolescence, who are struggling with their own identity and morals development, become especially susceptible to peer pressure and the need to "fit in." Education focused on the physiological effects of drug use and dependency provides students with the knowledge needed to make responsible choices regarding their own behavior.

To develop therapies that will help people, scientific research sometimes involves the inclusion of animals. This is true for studies to help treat cancer or infection, as well as studies that look for therapies for diseases that affect the brain, kidney, or other organs of the body. The inclusion of animals in research has historically played a pivotal role in the advancement of medical technology and illness prevention. The need for studies with animals is especially important in the area of drug abuse research. Many invaluable experiments could not be conducted with human volunteers, such as uncovering the mechanisms of drug addiction or the search for safe alternative treatment methods. Our society has said that use of animals for this purpose, of helping humanity and assuring a better future for our environment, is ethical—but only under certain strict conditions. Within the lessons in this book we explore these issues in response to the priorities set by the National Institutes of Health. There are two objectives: First, we want students to learn how science is helpful and second, we want to examine how science is conducted and overseen in order to assure that it is careful and ethical. Although the need to treat animals humanely is widely accepted and rigorously monitored, media exposés of animal mistreatment (e.g., in the food and cosmetics industries and by some medical researchers) have contributed to increased interest, especially among young people. This curriculum will provide a foundation for students to begin to formulate their own answers to ethical questions concerning the use of animals in medical research.

According to the most recent data from the U.S. Census Bureau (2000), there were approximately 49 million children and youth between the ages of 6 and 17 living in the United States. These young people are the future of our country and they are growing up during a period of explosive scientific growth, particularly in the areas of the neurosciences—including genetics, basic science, and medication development. The need for students to understand the value of the neurosciences and the damaging effects of illicit drug use, the mechanisms of addiction, and of the scientific and ethical basis of animal-based drug abuse research is critical.

10-LESSON UNIT GOALS AND OBJECTIVES

This Is Your Brain: Teaching About Neuroscience and Addiction Research focuses on the neurobiology and consequences of drug use and abuse, the positive role of drug abuse research in our society, and the ethical guidelines and practices for inclusion of animals in research. The unit meets an important ethical need in contemporary scientific education for young people and is designed to generate greater understanding and support for scientists and clinicians who rely on evidence derived from animal research studies.

Developed to cover key scientific and ethical content domains, this unit includes specific lesson plans for teachers, interactive learning materials for students, and companion materials for parents. The content complies fully with the National Research Council's *National Science Education Standards* and all existing ethics standards related to biomedical research.

Key Unit Learning Objectives

As a result of completing the 10-lesson unit, middle school science students will be able to:

1. describe the fundamental principles of the scientific method

2. identify the effects drug use and abuse have on the structure and function of various human body systems

3. develop an understanding of the social/family, biological, legal, academic, psychological/behavioral, and economic consequences of drug use and misuse

4. analyze risk factors associated with drug use and abuse

5. explain the structure and specialized function of the brain and neuron cells

6. demonstrate the process of message transfer between neuron cells

7. illustrate how drugs of abuse interfere with the communication process between neurons

8. define disease and formulate an explanation of why drug addiction is considered a brain disease

9. distinguish the difference between drug abuse and addiction

10. develop an understanding of the four different approaches to scientific research

11. analyze the potential harms/benefits of scientific research to animals and society

12. identify the historical contributions of animals in medical research

13. demonstrate an understanding of the following ethics terms: *autonomy, beneficence, compassion, justice, nonmaleficence,* and *veracity*

14. explain the role of an Institutional Animal Care and Use Committee (IACUC) and federal guidelines such as the Animal Welfare Act

15. describe the three Rs (reduce, refine, and replace) of animal inclusion in research and how researchers apply these principles

10-LESSON OVERVIEW

I. Development of a Teen Profile

Lesson 1: Who Is Chris?

Through class and small-group activities, students develop a profile of a teen named Chris, including his/her physical attributes, hobbies, goals, family and school life. Throughout the unit, the story of this teen unfolds as concepts pertaining to drug addiction, biomedical research, and animal inclusion in research are explored. Emphasis is placed on the process of the scientific method.

II. Overview of Brain Structure and Drug Addiction/Abuse

Lesson 2: Thinking Like a Scientist

Students explore the importance of the scientific method and its application both to historic scientific discoveries and our everyday lives. The story of Chris continues to unfold as students discover Chris has tested positive for drugs. Students analyze and categorize a variety of facts related to drug use. By using decision-making skills, students discuss and gain awareness of the many aspects of Chris's life that have been affected by the use of drugs.

Lesson 3: How Does the Brain Work?

Through exploration and discovery, students are introduced to the general regions and functions of the human brain. Neuron structure and purpose are explored, as well as how the brain sends and receives messages. Students build on this inquiry to address the importance of questions in gathering information to understand scientific processes.

Lesson 4: How Do Drugs Affect the Brain?

Students build on Lesson 3's exploration of neuron structure to understand how neurons transfer messages and how drug use and abuse negatively interferes with the function of the brain. Using critical-thinking skills, students analyze the risks associated with drug use and abuse, the science of addiction, and how drug use changes the normal functioning of human body systems.

III. How Science Is Helping Us Understand Drug Addiction

Lesson 5: How Science Is Helping

In this lesson, students explore the four categories of scientific research that are helping us understand drug abuse and addiction: biological, population-based, behavioral, and genetic research. Students analyze and sort a variety of research studies to develop an understanding of how knowledge derived from scientific research translates into clinical practice.

Lesson 6: Why Research Is Important

Students explore, analyze, and sort a variety of research studies. Emphasis is given to illustrations of the historical contributions animals have made in medical research. Students begin to consider some basic ethical issues surrounding the inclusion of both animals and humans in research.

IV. Thinking Like a Scientist: Research and Ethics

Lesson 7: What Is Ethics in Science?

Students investigate and clarify their understanding of the concept of ethics. They explore the meaning of six key ethics principles and how these principles might apply to hypothetical situations.

Lesson 8: Applying Ethics to Research

Students explore ethical considerations in research and focus on the specific example of the inclusion of animals in medical research. They work in small groups to analyze the potential harms and benefits to animals in two hypothetical research scenarios, with the goal of deciding which, if either, they would approve for study. They learn about the factors and guidelines that are important for a researcher to follow to ensure ethical treatment of animals.

Lesson 9: Ensuring the Ethical Conduct of Research

Researchers must comply with very strict rules when doing their scientific work. In this lesson, students learn about studies involving animals that have helped scientists, doctors, and veterinarians to develop new medical treatments, not only for humans but also for animals. Scientists look for ways to reduce, replace, and refine (the three Rs) the number of animals included in research. In this lesson, students explore the federal regulations (e.g., Animal Welfare Act) that are in place to protect animals, and discuss acceptable practices for their inclusion in research.

Lesson 10: Thinking about the Future of Research

Students explore the benefits and shortfalls of alternative research methods, and discuss when and how and these options might be incorporated to help achieve the three Rs. For example, brain imaging technologies and computer simulations may be an appropriate replacement to animal testing, and genetic DNA markers can help refine the extent to which animals are involved in a research project.

NATIONAL SCIENCE EDUCATION STANDARDS

T he National Research Council's *National Science Education Standards* (NSES) is the result of the productive collaboration of hundreds of teachers, parents, school administrators, curriculum developers, college professors, scientists, and government officials. Within a framework of developmentally appropriate levels of scientific and ethical knowledge and understanding, the standards emphasize an approach that is

- "hands on" (i.e., involving students in experiments and observations), and
- "minds on" (i.e., stimulating students to inquire, describe, ask questions, formulate and test explanations, and reason about conclusions logically).

Designed to promote "coordination, consistency, and coherence to the improvement of scientific education," the NSES serve as the criteria upon which local, state, and national programs judge the quality of their science education efforts. Not only do these standards identify what students "need to know, understand, and be able to do to be scientifically literate at different grade levels," they provide an educational framework within which teachers can create an effective learning environment that promotes the outstanding performance of all students. To help ensure effective and high-quality content, the Research and Ethics lessons were specifically developed to meet a number of these critical standards (see next page).

In addition, the lessons have been designed to encourage students to reflect on prior knowledge, explore and question new information, and make connections to their everyday lives and the real world in a more complex manner. Founded on the principles outlined in the Biological Sciences Curriculum Study (BSCS) 5E Instructional Model (see p. 12), this process helps students become self-directed learners and construct knowledge that is meaningful to them, an essential process in the cognitive development of learners.

NATIONAL SCIENCE EDUCATION STANDARDS MET IN UNIT

Content Standards: Grades 5–8	Lesson Correlation
Standard A: Science as Inquiry	
• Abilities necessary to do scientific inquiry	All lessons
• Understanding about scientific inquiry	All lessons
Standard B: Physical Science	
• Transfer of energy	3 and 4
Standard C: Life Science	
• Structure and function in living systems	2, 3, 4, 5, 6, 8, and 10
• Regulation and behavior	2, 3, 4, 5, 6, and 8
Standard E: Science and Technology	
• Abilities of technological design	8 and 10
• Understanding about science and technology	10
Standard F: Science in Personal and Social Perspectives	
• Personal health	2, 4, 5, 6, 7, 8, and 9
• Risks and benefits	2, 4, 5, 6, 7, 8, and 9
• Science and technology in society	2, 4, 5, 6, 8, 9, and 10
Standard G: History and Nature of Science	
• Science as a human endeavor	1, 2, 6, and 7
• Nature of science	2, 5, 6, 7, and 8

5E INSTRUCTIONAL MODEL

THE BIOLOGICAL SCIENCES CURRICULUM STUDY (BSCS) 5E INSTRUCTIONAL MODEL

Engage

These experiences mentally engage the students with an event or question. Engagement activities help students make connections with what they know and can do. During the engagement phase, the teacher can

- Create a need to know/create an interest
- Assess prior knowledge
- Focus on a problem/ask questions

Explore

Students work with one another to explore ideas through hands-on activities. Under the guidance of the teacher, students experience a common set of experiences that helps them clarify their own understanding of major concepts and skills. During the exploration phase, the students

- Investigate
- Develop awareness/practice skills
- Design, plan, build models, collect data
- Test predictions and form new predictions

Explain

Students explain their understanding of the concepts and processes they are learning. Teachers help students clarify their understanding and introduce information related to the concepts to be learned. During the explanation phase, teachers and students

- Clarify understanding
- Define concepts or terms
- Share understandings for feedback
- Listen critically to one another
- Form generalizations
- Refer to previous activities

Elaborate

These activities challenge students to apply what they have learned and extend their knowledge and skills. During the elaboration phase, students

- Build on their understanding of concepts
- Use knowledge of concepts to investigate further—extension
- Apply explanations and skills to new, but similar, situations
- Provide practice and reinforcement— application

Evaluate

Students assess their own knowledge, skills, and abilities. Evaluation activities also allow teachers to evaluate students' progress. During the evaluation phase, students

- Draw conclusions using evidence from previous experiences
- Demonstrate an understanding or knowledge of a concept or skill

LESSON FORMAT AND TIMELINE

This Is Your Brain contains all the necessary teacher materials—concepts and overviews, standards and objectives, procedures, group and hands-on activities, material supply list, time management estimates, color reproducibles, answer keys, resource lists[1], optional extensions, daily assessments, several final unit assessment options, a computer-based Mouse Maze knowledge assessment game, and a parent letter describing the course—in a user-friendly kit.

To ensure an optimal learning experience for the students, it is recommended teachers complete the *entire* 10-day curricular unit in sequence, because each lesson builds on the previous one. Each lesson is designed for a minimum 45-minute class and includes optional extension activities that can be incorporated if desired.

A PowerPoint of the 20 transparencies in these lessons is available for your convenience. Visit *www.nsta.org/publications/press/extras/brain.aspx* to download.

---

1 Numerous website links are provided throughout the unit for teacher, student, and parent use. Please note that the internet is a continually changing environment and these URLs are subject to change without notice.

FLEXIBILITY OF UNIT

Teachers can easily adapt the lessons and activities to suit a variety of grade levels and learning abilities. The suggested timeline can be modified to suit the needs of varying school schedules, curriculum maps, and individual teaching styles.

ASSESSMENT OPTIONS

*T*his Is Your Brain includes a summative unit test (multiple-choice, short-answer, and essay questions) for use by students in paper format, as well as a multiple-choice assessment in the form of an interactive, computer-based Mouse Maze game. The unit also provides several authentic assessment options, including: creating a board game, producing an antidrug commercial, developing a school or class newspaper, and designing an informative poster. Assessment options can be used throughout the unit or at its completion to ensure effective learning.

MATERIAL SUPPLY LIST

LESSON	ACTIVITY ONE	ACTIVITY TWO	ACTIVITY THREE	ACTIVITY FOUR
1	overhead projector Transparency A Worksheet 1	Worksheet 1	overhead projector Transparency B Lesson 1 homework extension three-ring binder or folder three-hole punch	
2	overhead projector Transparencies B, C, & D Worksheets 2 & 3	Worksheet 4 Transparencies E & F	overhead projector Transparency F Steps of the scientific method bookmark Worksheet 2 Lesson 2 homework extension	
3	overhead projector Transparencies F & G Worksheet 5	overhead projector Transparencies H & I Worksheets 6, 7, & 8	overhead projector transparencies H & J Lesson 3 homework extension	
4	overhead projector Transparency J colored pencils or crayons Worksheet 9 20 small objects of 2 different colors	Worksheet 9 internet access Transparency I (optional)	overhead projector Transparency K Lesson 4 homework extension	
5	overhead projector Transparencies K & L research labels Worksheet 10	Worksheet 11	overhead projector Transparency M Lesson 5 homework extension	

LESSON	ACTIVITY ONE	ACTIVITY TWO	ACTIVITY THREE	ACTIVITY FOUR
6	overhead projector Transparency M 5 large sheets of poster paper Worksheet 12 animal cards (1 sheet per group) scissors tape	student glossary	overhead projector Transparency N Lesson 6 homework extension internet access	
7	overhead projector Transparency N Worksheet 13	overhead projector Transparency O Worksheets 13 & 14	Worksheet 13	overhead projector Transparency P Lesson 7 homework extension
8	overhead projector Transparency P scenario #1 handout scenario #2 handout Worksheet 15	scenario #1 handout scenario #2 handout Worksheet 15	overhead projector Transparency Q Lesson 7 honmework extension	
9	overhead projector Transparency Q Worksheet 16	Talking Points handout	Worksheet 17	overhead projector Transparency R Worksheet 17
10	overhead projector Transparency S 1 set of three Rs research cards per group Worksheet 18	overhead projector Transparency R Worksheet 19		

VOCABULARY TERMS INTRODUCED IN EACH LESSON

Lesson 1:
characteristic
hypothesis
profile

Lesson 2:
brain
cocaine
data
drug
ecstasy
experiment
heroin
inhalant
illicit
licit
LSD
marijuana
neuron
nicotine
scientific method
steroid

Lesson 3:
axon
axon terminals
brainstem
cell body
cerebellum
chemical message
dendrites
dopamine
electrical message
frontal lobe
GABA
myelin
neurotransmitter
occipital lobe
parietal lobe
receptors
serotonin
soma
synapse
synaptic gap
temporal lobe

Lesson 4:
abuse
addiction
chronic
craving
disease
drug abuse
drugs of abuse
limbic system
reward
tolerance

Lesson 5:
behavioral research
biological research
biomedical research
control group
experiment
genetic research
population-based research
research
THC

Lesson 6:

model

Lesson 7:

autonomy
beneficence
bioethics
compassion
ethics
explicit
implicit
justice
opinion
morals
nonmaleficence
values

Lesson 8:

protocol
Institutional Animal Care and
    Use Committee

Lesson 9:

animal welfare
Animal Welfare Act (AWA)
United States Department of
    Agriculture (USDA)

Lesson 10:

anesthesia
blood work
cell culture
chemical simulation
computer simulations
genetic alterations
genetic markers
magnetic resonance imaging
    (MRI)
mathematical modeling
mechanical simulation
non-mammalian models
PET scan
reduce
refine
replace
Three Rs
tissue culture

SECTION TWO

LESSONS

LESSON 1
WHO IS CHRIS?

Teacher Lesson Plan: Overview and Concepts

In this lesson, students are introduced to a teenager named Chris, whose story unfolds with each subsequent lesson. The only known fact about Chris is contained in a headline, challenging students to begin thinking like scientists about this hypothetical character.

Chris could be a male or female, tall or short, a genius or an average student. It is up to your students to develop a profile and hypothesis about Chris. As they define Chris's physical and social attributes, students come to know Chris and become invested in the outcome of this teenager's future. Through class and small-group activities, students explore the scientific method and lay the groundwork for developing a hypothesis about Chris's problem.

Activities include selecting goals, appearance, and personality characteristics that will be expressed in written and visual formats. The homework for Lesson 1 becomes the cover page of a project portfolio that students will work on throughout the unit.

Objectives

After completing this lesson, students will be able to

- demonstrate an understanding of the term *hypothesis*;
- work cooperatively in small groups to develop and write a biographical profile;
- apply oral presentation and decision-making skills;
- create a visual image based on a written profile; and
- formulate appropriate questions and answers to further scientific inquiry.

National Standards Met in Lesson 1

National Science Education Standards

Standard A: Science as Inquiry

- Abilities necessary to do scientific inquiry
- Understanding about scientific inquiry

Standard G: History and Nature of Science

- Science as a human endeavor

National Council of Teachers of English

- Students employ a wide range of strategies as they write and use different writing process elements appropriately to communicate with different audiences for a variety of purposes. (standard 5)

- Students conduct research on issues and interests by generating ideas and questions and by posing problems. They gather, evaluate, and synthesize data from a variety of sources (e.g., print and nonprint texts, artifacts, people) to communicate their discoveries in ways that suit their purpose and audience. (standard 7)

Consortium of National Arts Education Associations—Visual Arts

- Students intentionally take advantage of the qualities and characteristics of art media, techniques, and processes to enhance communication of their experiences and ideas. (standard 1)

- Students select and use the qualities of structures and functions of art to improve communication of their ideas. (standard 2)

ACTIVITY ONE
INTRODUCE CHRIS

Time needed for completion:

20 minutes

MATERIALS

For the class:

- Overhead projector
- Transparency A

For each student:

- Worksheet 1

Procedure

1. Display Transparency A: *Chris Collapses in Gym!* (p. 29) Ask, "What do you think happened to Chris?"

2. Give students time to think about the headline. Then explain that in future lessons students will be developing a hypothesis about Chris. Review the definition of *hypothesis* (refer to Unit Glossary, p. 216). Answers could be "educated guess" or "assumption." Explain that *hypothesis* comes from a Greek word that means "to suppose." A hypothesis is a prediction about what will occur in a specific situation. Stress to students that a hypothesis is not a question. Scientists often ask many questions before they form a hypothesis.

3. Distribute Worksheet 1: *Who Is Chris?* (pp. 32–33) to each student.

4. In order to develop this hypothesis, students will have to answer the question, "Who is Chris?" Each student will be completing a profile about Chris. Explain that a written profile of a person is a set of biographical details. There are no right or wrong answers; students can create profiles based on their own perceptions and ideas.

5. To demonstrate how to complete the worksheet, have the class decide Chris's age. (Select from an appropriate range between 12 and 16.)

6. Have students work in small groups (three or four) to complete all of Worksheet 1: Part A. Each group should agree upon the characteristics assigned to Chris's profile. Encourage students to make Chris as "real" as possible. These small groups will work together throughout the unit.

ACTIVITY TWO
PRESENT CHRIS'S PROFILE TO THE CLASS

1

Time needed for completion:
20 minutes

MATERIALS

For each student:

- Worksheet 1

Procedure

1. Have each group select a representative to present to the class approximately five characteristics from their profile of Chris that was completed in Worksheet 1: Part A.

2. Encourage students to elaborate on their descriptions and why the characteristics were selected for Chris's profile.

   Questions for additional discussion could include*:

   - Are any of Chris's characteristics like those of a character from a book you have read?

   - How is Chris similar to you?

   - How is Chris different from you?

   *Please see *optional extensions* (at the end of this lesson) for ways to bring Chris to life in your classroom.

3. Have students complete Worksheet 1: Part B independently.

1

ACTIVITY THREE

CLOSING TEASER— TOMORROW'S HEADLINE

Time needed for completion:

5 minutes

MATERIALS

For the class:

- Overhead projector
- Transparency B

For each student:

- Lesson 1 homework extension
- 3-ring binder or folder

Student Project Portfolio

For this unit, students will compile a project portfolio of the activities and homework extensions for every lesson. The pages involved can be distributed all at once and either stapled or kept in a binder or folder; or they may be handed out as needed. If you are going to provide binders or folders for the students, hand them out on the first day of the unit. Also give each student a copy of the Student Glossary (photocopy pages 251–259) on the first day, which can be kept at the back of the binder or folder for easy reference.

Procedure

1. After completing the activities, display Transparency B: Chris Treated at Emergency Room without providing additional details about Chris.

2. Hand out Homework Extension: Project Portfolio Cover Page for students to complete before the next lesson. This will be the first page of the student project portfolio. Have each student draw a portrait of Chris based on the profile developed in the small-group activity. Encourage student individuality in selecting materials to create the portrait of Chris. This could be a realistic portrait, abstract, 3-D, cartoon, stick figure, or other art form.

Optional Extensions

- Have students write five factual statements about Chris based on Transparencies A and B.

- Have groups create a medical profile for Chris, following the medical form that is used by the school district. This could include facts gathered during a routine exam such as pulse, blood pressure,

1 ACTIVITY THREE
CLOSING TEASER—TOMORROW'S HEADLINE

temperature, blood type, history of illnesses, medical treatments, immunizations, and medications.

- Have the class create a three-dimensional (3-D) Chris, using a mannequin or scarecrow format, to keep in the classroom during the course of the unit or to display in the school to encourage all-school interest in Chris's story. Groups could take turns creating Chris's wardrobe and accessories.

- Assign each group to bring to class an object that represents Chris's life: favorite book, article of clothing, letter written by or to Chris, family portrait, report card, and so on. Gather the items in a suitcase or box that is labeled

"Chris." Include this extension throughout the unit to encourage creativity and individual expression as students define Chris and identify with Chris's condition.

Resources for Further Exploration

Check your school library for books about scientific inquiry.

Parent Letter

Prior to beginning the first lesson, the parent/guardian letter may be sent home describing what the students will be learning throughout the unit. Space is provided for questions and/or comments as well as the parent/guardian signature.

WHO IS CHRIS?
TRANSPARENCY B

Name_____ Date_____

PROJECT PORTFOLIO COVER PAGE

HOMEWORK EXTENSION

Directions: Use the information from Worksheet 1 to create a portrait of Chris.

Name_____ Date_____

WHO IS CHRIS?
WORKSHEET 1

Part A—Group

Fill in the blanks to create a profile of Chris.

Age _____

Gender _____

Height_____

Hair color _____

Favorite book _____

Favorite song _____

Grade point average _____

Favorite food _____

Favorite sport _____

Favorite class _____

Favorite color _____

WHO IS CHRIS?
WORKSHEET 1 (CONTINUED)

Part A—Group (continued)

Complete these sentences.

Chris's personality can be described as _____

_____

Chris's dream job would be _____

_____

If Chris could change one thing about the world, it would be _____

_____

The best gift Chris ever received was _____

_____

Chris's family includes _____

_____

Chris's favorite science activity or experiment is _____

_____

Part B—On Your Own

Fill in the blanks to complete the activity.

A set of biographical details about a person is called a _____

The Greek word for "to suppose" is _____

What question would you like to ask Chris that has not already been answered in class? _____

_____

LESSON 2
THINKING LIKE A SCIENTIST

Teacher Lesson Plan: Overview and Concepts

In this lesson, students explore the importance of the scientific method and how it applies to both historic scientific discoveries and our everyday lives. The story of Chris continues to unfold as students follow the step-by-step process of the scientific method to develop a hypothesis about Chris.

As students learn that Chris has tested positive for drugs, they think like scientists about the problem being presented. Using analytic and decision-making skills, students discuss and gain awareness of the many aspects of Chris's life that have been affected by drug abuse.

Students collect and analyze data to create and interpret graphs and to formulate a question for Chris's doctor to explore during a medical exam. Students also provide an example of how the scientific method applies to their daily lives.

Objectives

After completing this lesson, students will be able to

- work cooperatively in small groups to identify, analyze, and categorize information to draw valid conclusions;

- organize and interpret research data to draw conclusions;

- identify the effects of drug use and abuse on the structure and function of human body systems;

- develop an understanding of the social/family, biological, legal, academic, and psychological/behavioral consequences of drug use and misuse;

- develop thoughtful questions that reflect the scientific inquiry process; and

- analyze risk factors associated with drug use and abuse.

National Standards Met in Lesson 2

National Science Education Standards

Standard A: Science as Inquiry

- Abilities necessary to do scientific inquiry
- Understanding about scientific inquiry

Standard C: Life Science

- Structure and function in living systems
- Regulation and behavior

Standard F: Science in Personal and Social Perspectives

- Personal health
- Risks and benefits
- Science and technology in society

Standard G: History and Nature of Science

- Science as a human endeavor
- Nature of science

National Health Education Standards

- Students explain the relationship between positive health behaviors and the prevention of injury, illness, disease, and premature death.

- Students describe the interrelationship of mental, emotional, social, and physical health during adolescence.
- Students analyze how environment and personal health are interrelated.

(standard 1 — all of the above)

National Council of Teachers of Mathematics

- Students formulate questions that can be addressed with data and collect, organize, and display relevant data to answer questions.
- Students develop and evaluate inferences and predictions that are based on data.
- Students create and use representations to organize, record, and communicate mathematical ideas.

National Council of Teachers of English

- Students employ a wide range of strategies as they write and use different writing process elements appropriately to communicate with different audiences for a variety of purposes. (standard 5)

ACTIVITY ONE
THE SCIENTIFIC METHOD

Time needed for completion:

15 minutes

MATERIALS

For the class:

- Overhead projector
- Transparency B
- Transparency C
- Transparency D

For each student:

- Worksheet 2
- Worksheet 3

Procedure

1. At the start of class, show Transparency B: *Chris Treated at Emergency Room* from Lesson 1.

2. Display Transparency C: *Chris Tests Positive for Drugs!* Give students time to speculate about the headline. Ask, "Does this information change the kinds of questions you wanted to ask Chris?" Explain that scientists often have to revise their questions as they work through problems using the scientific method.

3. Distribute Worksheet 2: *The Scientific Method* (p. 47) to each student and display Transparency D: *The Scientific Method*. (Note: Worksheet 2 may be used as a guide if the school or district uses a different version of the scientific method.) Discuss, in general, the steps of the scientific method and explain to students that each step will continue to be explored throughout the unit. Have students share the questions they wanted to ask Chris in Worksheet 1: Part B (from Lesson 1) and after viewing Transparency C: *Chris Tests Positive for Drugs!* Ask, "In what step of the scientific method does your question fall?"

4. Discuss the importance and some outcomes of inquiry in science, such as:

Question	Discovery/Invention
How do birds fly?	Airplane
Why does the sun rise and set?	Earth is round and rotates
Is there a faster method to calculate?	Abacus, then computers

5. Distribute Worksheet 3: *Using the Scientific Method* (pp. 48–49) to students in their small groups. Assign or allow groups to choose a question around which they will design an experiment. (This activity can also be conducted as a class.) Suggestions for questions are:

    - How often do your classmates wear the color red?

    - How does the grade point average in your classroom compare to the school average?

    - Who do middle school students consider to be the most important role model in their lives?

6. Students follow the guide on their worksheets and fill in the space for each step of the scientific process as they plan how they would conduct an experiment to answer their question. Provide time for groups to present their plans to the class.

ACTIVITY TWO

COLLECTING, ORGANIZING, AND INTERPRETING DATA

Time needed for completion:

25 minutes

MATERIALS

For the class:

- Overhead projector
- Transparency E
- Transparency F

For each student:

- Worksheet 4

Procedure

1. Distribute Worksheet 4: *Looking at the Facts* (pp. 50 – 51) and explain that the categories listed in the table are aspects of life that are often affected by a person's drug use and misuse.

2. Explain that in collecting information, scientists must sort and categorize data in order to understand the problem or question. For example, a scientist conducting a study on rain patterns will collect facts, or raw data, and then categorize this information according to hours, days, weeks, temperature, location, and wind conditions.

3. Explain that the table on Worksheet 4 is similar to the way scientists organize data. The data in this table might be used to develop a hypothesis about Chris.

4. Display Transparency E: *Types of Graphs* and provide students with a brief overview of simple graphs. Discuss the two types of graphs (bar and line), the purpose for each, what the *x* and *y* axes represent, and the importance of labels for each axis and the graph overall. Describe sample data that might be represented by these graphs such as:

   - The number of hours of television each person or group watched on a Saturday
   - The amount of money spent by five students on lunch during one week
   - The number of students in a classroom who wore the color red over a five-day period

5. Discuss how the graph might be labeled (*x* and *y* axes, graph title) depending on the focus selected.

6. Allow students to work in their small groups to complete Worksheet 4, parts A–C. Have each group take a turn presenting their graph to the class. A representative from each group should describe the group's decision-making process in interpreting the data and creating their graph.

7. Have students independently write a new question to ask Chris as requested in Worksheet 4: Part D.

ACTIVITY THREE

CLOSING TEASER— TOMORROW'S HEADLINE

Time needed for completion:

5 minutes

MATERIALS

For the class:

- Overhead projector
- Transparency F

For each student:

- Steps of the Scientific Method bookmark (see p. 54 for printable page)
- Worksheet 2
- Lesson 2 homework extension

Procedure

1. Distribute the Steps of the Scientific Method bookmark to each student. (Laminate if available; other options are to attach the paper copy to cardboard or print directly to cardstock. Consider punching a small hole in the corner of the bookmark so it can be kept in the student's project portfolio.)

2. After completing the other activities, assign the homework extension, then display Transparency F: *What's Wrong, Chris?* without providing additional details.

Homework Extension: Using the Facts

Hand out Homework Extension: *Using the Facts* (p. 52) for students to complete before the next lesson.

Optional Extensions

- Have students interview a doctor or other healthcare professional (school nurse, psychologist, and so on) about the importance of using the scientific method in their work. Students should prepare a list of questions in advance.

- Have students research one of the historic scientific questions listed below. Using Worksheet 2: *The Scientific Method* (p. 47), have students write out the steps they would have followed to answer the question.

  - What causes yellow fever?
  - What causes diabetes?
  - What causes polio?
  - What causes AIDS?

ACTIVITY THREE
CLOSING TEASER—TOMORROW'S HEADLINE

Resources for Further Exploration

Websites

NIDA Info Facts, National Institute on Drug Abuse,
    National Institutes of Health
    www.drugabuse.gov/NIDAHome.html
The Centers for Disease Control and Prevention
    www.cdc.gov

Books

Kramer, S. P. 1987. *How to think like a scientist:
    Answering questions by the scientific method.*
    New York: HarperCollins.

THINKING LIKE A SCIENTIST
TRANSPARENCY D: THE SCIENTIFIC METHOD

1. State the problem
(What are you wondering about?)

- Write the question(s) that the experiment will try to answer

2. Collect information
(What do you already know? What can you find out?)

- Look for information about your question before you begin your experiment
- How have others looked at this question?

3. Form a hypothesis
(Based on what you know, what do you think will happen?)

- Predict what the result of your experiment will show
- Try to write your hypothesis as an "if/then" statement

4. Test your hypothesis
(How will you conduct your experiment?)

- Describe the steps of your experiment
- What materials will you need?

5. Observe and record your results
(What happened in your experiment?)

- Organize and record the data you collected

6. Draw a conclusion
(Did you support your hypothesis? What new questions do you have?)

- State whether your hypothesis was supported
- Share your conclusion with others

THINKING LIKE A SCIENTIST
TRANSPARENCY E: TYPES OF GRAPHS

BAR GRAPHS

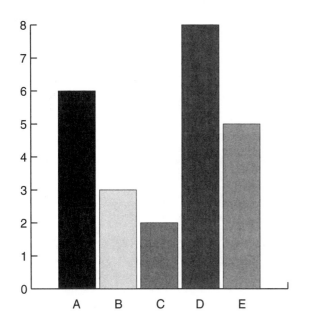

LINE GRAPHS

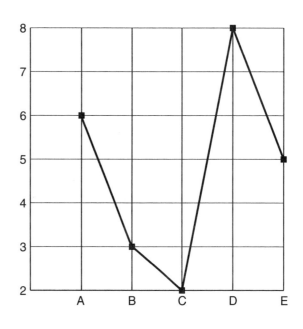

THE SCIENTIFIC METHOD
WORKSHEET 2

1. State the problem

What are you wondering about?

- Write the question(s) that the experiment will try to answer

2. Collect information

What do you already know?
What can you find out?

- Look for information about your question before you start your experiment

- How have others looked at this question?

- What will you measure?

3. Form a hypothesis

Based on what you know, what do you think will happen?

- Predict what the result of your experiment will show.

- Try to write your hypothesis as an "if/then" statement

4. Test your hypothesis

How will you conduct your experiment?

- Describe the steps of your experiment

- What materials will you need?

5. Observe and record your results

What happened in your experiment?

- Organize and record the data you collected

6. Draw a conclusion

Did you support your hypothesis?
What new questions do you have?

- State whether your hypothesis was supported

- Share your conclusion with others

Name_____ Date_____

USING THE SCIENTIFIC METHOD
WORKSHEET 3 Student responses will vary based on scientific question.

1. State the problem
(What are you wondering about?)

2. Collect information
(What do you already know? What can you find out?)

3. Form a hypothesis
(Based on what you know, what do you think will happen?)

USING THE SCIENTIFIC METHOD
WORKSHEET 3 (CONTINUED)

4. Test your hypothesis
(How will you conduct your experiment?)

5. Observe and record your results
(What happened in your experiment?)

6. Draw a conclusion
(Did you support your hypothesis? What new questions do you have?)

Name_____ Date_____

LOOKING AT THE FACTS
WORKSHEET 4

Part A

The categories in the table below show the areas of a person's life that can be negatively impacted by using drugs. Look at the table below and discuss the meaning of the data with your group.

A group of 10 middle school students were surveyed on how they thought drug use might most negatively impact areas of their lives. Students rated each topic with "1" indicating the area they felt would be least negatively impacted, "6" most negatively impacted.

Category	Student ID number										Total
	A	B	C	D	E	F	G	H	I	J	
	Ratings										
Friendships	5	6	4	4	1	5	2	1	1	6	35
Grades	2	3	3	6	2	4	6	4	4	5	39
Family relationships	6	5	5	5	3	6	5	6	2	4	47
Sports performance	1	1	2	1	6	1	1	3	3	3	22
Health	4	2	6	3	4	2	3	2	5	2	33
Wake/sleep cycles (e.g., falling asleep in class, difficulty waking up in the morning)	3	4	1	2	5	3	4	5	6	1	34

1 = least impact; 6 = most impact

Part B

Decide what type of a graph you need.

- Line Graphs are used to track changes over time.

- Bar Graphs compare data between different groups or changes over time.

LOOKING AT THE FACTS
WORKSHEET 4 (CONTINUED)

Part C
Create your graph.

Many graphs have an *x* axis and a *y* axis. The *x* axis (a horizontal line) usually has numbers or labels for what is being measured, and the *y* axis (a vertical line) has numbers for the amount of the things being measured.

In the above table, what is being measured or compared? (*x* axis)

_____

Where can you find the measurement or comparison information? (*y* axis)

_____

Using graph paper, create a graph based on the information in the table on p. 50. Create your graph using a right angle with an *x* and *y* axis, labels for each axis, and points to represent your data. Also, don't forget to create a title for your graph.

Part D
Based on the information discussed in this lesson, what is a new question that Chris's doctor might ask Chris?

_____

_____

_____

_____

_____

_____

_____

Name_____ Date_____

USING THE FACTS
HOMEWORK EXTENSION

The graph below is based on a survey conducted by The National Household Survey on Drug Abuse in 2000. Students were asked about their use of cigarettes and illicit (illegal) drugs the month before the survey was taken. This graph shows the results of that survey.

Academic Performance and Use of Cigarettes or Illicit Drugs by Youth Ages 12–17

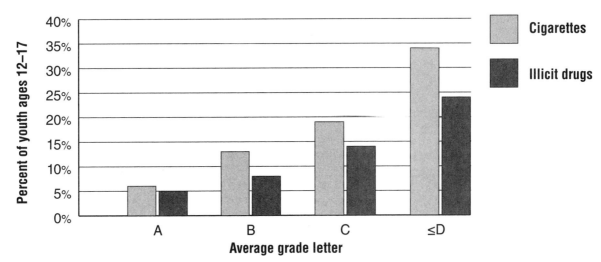

1. How is each axis labeled in the graph above? Explain why.

   x axis - _____
   _____
   _____
   y axis - _____
   _____
   _____

   How old were the students being studied?
   _____

2. What are the percentages by grade letter of reported illicit drug use in the students?

   A = _____%   B = _____%

   C = _____%   ≤ D = _____%

USING THE FACTS
HOMEWORK EXTENSION (CONTINUED)

3. Students with grades of ≤ D used about _____ % more cigarettes and about _____ % more illicit drugs than students with a grade of A.

4. Write one conclusion based on the facts in this graph.

    _____
    _____
    _____
    _____

The National Institute on Drug Abuse (NIDA) collected the following data in 1995 and again in 2003 on eigth-grade students who used illicit drugs:

Illicit Drug	Percent in 1995	Percent in 2003
Marijuana/Hashish	19.9	17.5
Inhalants	21.6	15.8
Hallucinogens and LSD	9.6	6.1
Cocaine and Crack Cocaine	6.9	6.1
Heroin	2.3	1.6
Steroids	2.0	2.5

5. Using graph paper (or the back of this page), draw a graph to show the difference in the percentage of drug use between 1995 and 2003. (Remember to include a title, axis labels, and measurements on your graph.)

6. Develop a one-sentence conclusion based on the facts presented in your graph.

    _____
    _____
    _____
    _____
    _____
    _____
    _____

7. Write one question that you think the doctor will ask Chris based on the information provided.

    _____
    _____
    _____
    _____
    _____
    _____
    _____
    _____
    _____
    _____

USING THE FACTS
HOMEWORK EXTENSION (CONTINUED)

STEPS OF THE SCIENTIFIC METHOD

 1. State the problem

 2. Collect information

 3. Form a hypothesis

 4. Test your hypothesis

 5. Record your results

 6. Draw a conclusion

STEPS OF THE SCIENTIFIC METHOD

 1. State the problem

 2. Collect information

 3. Form a hypothesis

 4. Test your hypothesis

 5. Record your results

 6. Draw a conclusion

STEPS OF THE SCIENTIFIC METHOD

 1. State the problem

 2. Collect information

 3. Form a hypothesis

 4. Test your hypothesis

 5. Record your results

 6. Draw a conclusion

STEPS OF THE SCIENTIFIC METHOD

 1. State the problem

 2. Collect information

 3. Form a hypothesis

 4. Test your hypothesis

 5. Record your results

 6. Draw a conclusion

LESSON 3

HOW DOES THE BRAIN WORK?

Teacher Lesson Plan: Overview and Concepts

In this lesson, students continue the process of inquiry as they follow the steps of the scientific method. Through exploration and discovery, they identify regions and functions of the brain and explore how the brain sends and receives messages. In Lesson 2, students formulated questions about Chris. The activities in Lesson 3 will help the students recognize that carefully developed questions guide the scientific process.

Students will be introduced to the general functions of the human brain. Neuron structure and activity will be explored through the scientific processes of observation and recording information. Students, in turn, will interpret their records to complete the homework assignment. This lesson establishes the background knowledge necessary to understand how neurons work, the sequence of neuron message transmission, and how drugs interfere with the function and health of the brain.

Objectives

After completing this lesson, students will be able to

- develop and identify questions for further scientific inquiry;
- identify and interpret scientific information regarding the structure and specialized function of the brain and neuron cells;
- demonstrate the communication process between neurons;
- explain the transfer of energy that occurs when a message is delivered between neurons; and
- describe the purpose of neurotransmitters.

National Standards Met in Lesson 3

National Science Education Standards

Standard A: Science as Inquiry

- Abilities necessary to do scientific inquiry
- Understanding about scientific inquiry

Standard B: Physical Science

- Transfer of energy

Standard C: Life Science

- Structure and function in living systems
- Regulation and behavior

National Council of Teachers of English

- Students apply a wide range of strategies to comprehend, interpret, evaluate, and appreciate texts. They draw on their prior experience, their interactions with other readers and writers, their knowledge of word meaning and of other texts, their word identification strategies, and their understanding of textual features (e.g., sound-letter correspondence, sentence structure, context, and graphics). (standard 3)

- Students participate as knowledgeable, reflective, creative, and critical members of a variety of literacy communities. (standard 11)

- Students use spoken, written, and visual language to accomplish their own purposes (e.g., for learning, enjoyment, persuasion, and the exchange of information). (standard 12)

ACTIVITY ONE
FUNCTIONS OF THE BRAIN

Time needed for completion:
20 minutes

MATERIALS

For the class:
- Overhead projector
- Transparency F
- Transparency G

For each student:
- Worksheet 5

Procedure

1. Show Transparency F: *What's Wrong, Chris?* from Lesson 2. Have students share the questions they developed in Lesson 2's Home-work Extension: *Using the Facts,* (number 4: *Write one question that you think the doctor might ask Chris*).

2. After student discussion, explain that all questions are important ways of gathering information necessary to help gain an understanding of a problem or question. These questions lead toward the development of a hypothesis.

3. Explain that because Chris tested positive for drugs, the doctor will want to ask Chris more questions. By asking questions, the doctor will be better able to help Chris. Also, by asking questions, scientists are able to learn how the brain functions and how drugs affect the brain.

4. Display Transparency G: *Functions of the Brain.*

5. Share with students these facts about the human brain:

   - The brain feels like soft butter.
   - If all the wrinkles in your brain were ironed out, it would measure about 2.5 square feet or a little bigger than a student desk.
   - The brain uses 20% of the body's oxygen supply and 20% to 30% of the body's energy.
   - At birth the brain weighs about 1 pound and contains most of the nerve cells or neurons you have for your life.
   - By age 6, the brain's weight triples because of growth, branching, and increased connections among neurons.

ACTIVITY ONE
FUNCTIONS OF THE BRAIN

6. Ask students to perform an activity that does not require them to use their brain. Give students time to explore this *impossible* task. Emphasize that the brain regulates *all* human physiological, behavioral, and emotional functions.

7. Explain that the brain is the command center of the body and different areas of the brain process different kinds of information. Each part of the brain specializes in a specific kind of task.

8. Give each student a copy of Worksheet 5: *Functions of the Brain* (p. 68). Explain to students that Transparency G and Worksheet 5 show *general* areas of the brain and their functions (thought, speech, smell, breathing, hearing, memory, sight, sensation, perception, and coordination).

9. Discuss the function of each area of the brain using Transparency G and have students record each brain function on the appropriate line of Worksheet 5. Tell students that this worksheet can serve as a reference page in their project portfolios (they will need this information for other activities).

10. Explain that most brain functions involve several areas of the brain and some of these areas are deep inside the brain. The brain accomplishes all of these functions by sending and receiving messages through neurons. This is complex and occurs at very high speeds.

ACTIVITY TWO

HOW DOES THE BRAIN SEND AND RECEIVE MESSAGES?

Time needed for completion:

20 minutes

MATERIALS

For the class:

- Overhead projector
- Transparency H
- Transparency I

For each student:

- Worksheet 6
- Worksheet 7
- Worksheet 8

Procedure

1. Explain to students that in the last activity, they learned that:
   - the brain regulates all human physiological, behavioral, and emotional functions, and
   - the brain performs all of its functions by sending and receiving messages through neurons.

2. Discuss with the class what you know so far:
   - Chris collapsed
   - Chris tested positive for drugs
   - The brain regulates all body functions

3. Ask the class to use the information they have to form a statement about what has happened to Chris. Then ask what new questions they have. Resulting questions could be similar to: "How does the brain communicate messages?" and "How do drugs affect communication in the brain?"

4. Show a simple video of an electrical neural impulse and the communication between two neurons. (Example: *www.youtube.com/watch?v=-SHBnExxub8*). After students watch the video, ask them to explain how they think these neurons communicate with each other. After a brief discussion, show Transparency H: *How a Neuron Sends a Message* and distribute Worksheet 6: *How a Neuron Sends a Message* (p. 69). Describe the name and role of each part of the neuron. Have students fill in the labels on their worksheets according to the transparency. Instruct students to keep this worksheet for reference in their project portfolios (they will need this information for other activities).

ACTIVITY TWO
HOW DOES THE BRAIN SEND AND RECEIVE MESSAGES?

5. Give an overview of how messages are sent along a neuron from the soma down the axon to the axon terminal. This electrical message is then converted to a chemical message that crosses the synapse and comes into contact with a receptor on another neuron. This chemical message is converted back into an electrical message and the process begins again. A typical brain cell can have 1,000 to 10,000 connections to other brain cells. Imagine 10,000 strings reaching from one cell. Multiply that by 100,000,000,000 neurons in each brain!

Neuron Study Options

- If individual computers are available, have students perform the online exercise to build two neurons at *http://learn. genetics.utah.edu/content/addiction/reward/ madneuron.html*.

- If computers are not available, use paper manipulatives found on Worksheet 7: *Create a Neuron* (p. 70). Distribute two worksheets to each student. Instruct students to cut out the shapes, color and label, and assemble a set of two neurons.

6. Discuss with students a simple model of a neuron that they have with them all the time—their hands and arms. Have students describe how hands and arms resemble a neuron (fingers/dendrites; hand/soma; arm/axon; elbow/axon terminal). Also how a sleeve resembles a myelin sheath, and how placing a finger from the other hand at the elbow represents the synapse.

7. Show Transparency I: *Sending the Message Across the Synapse* and distribute Worksheet 8: *Sending the Message Across the Synapse* (p. 71).

8. Explain that messages could not be sent and received without a neurotransmitter. Describe each stage of message transfer across the synapse, as shown on the transparency. Have students fill in the blanks on their worksheets with the keywords included on the transparency (they will need this information for other activities). Give students the link—or show on an overhead—to this animated slide show with voiceover showing how neurons function and communicate:

http://learn.genetics.utah.edu/content/addiction/ reward/neurontalk.html

9. (Choosing the "without voiceover" option for the slide show provides an opportunity to discuss each step of the neuron communication process with students.) Explain that there are many types of neurotransmitters in the human body. Each neurotransmitter has an important role in a neural system that helps the body to function.

10. Explain that the brain has billions of neurons that send, receive, and store messages that are ultimately sent to other places in the brain and to the muscles and organs of the body. These messages make everything we

do possible: moving, breathing, digesting, thinking, experiencing sensations (heat, taste, cold, pain), and feeling emotions.

11. Have students provide some examples of the types of messages their brains are sending and receiving as they sit at their desks.

12. Tell students that in the next lesson they will be demonstrating how neurons send messages and how drugs interfere with the sending and receiving of messages between neurons.

ACTIVITY THREE

CLOSING TEASER— TOMORROW'S HEADLINE

Time needed for completion:

5 minutes

MATERIALS

For the class:

- Overhead projector
- Transparency J
- Transparency H

For each student:

- Lesson 3 homework extension

Procedure

After completing the other activities, display Transparency J: *How Have Drugs Affected Chris's Brain?* without providing additional details.

Homework Extension: The Brain Game Crossword Puzzle

Hand out Homework Extension: *The Brain Game Crossword Puzzle* (p. 72) for completion before the next lesson.

Optional Extensions

- Give each student an extra copy of Worksheet 5: *Functions of the Brain* to cut into puzzle pieces. Have students put the pieces in an envelope and pass to another student to reassemble.

- Have small groups brainstorm to create brain models using materials readily available in the classroom or visit *http://faculty. washington.edu/chudler/chsense.html* for guidelines on creating brain models from clay and gelatin or from one of the brain "recipes" included on the site.

- Assign groups of students to write brain messages on index cards to be arranged into a poem and shared with the class.

- Display Transparency H: *How a Neuron Sends a Message* for one minute. Remove the transparency and have students draw a neuron from memory. Repeat the process a few times. Discuss all the brain functions that are involved in this exercise.

- Have students visit *http://faculty.washington.edu/chudler/chsense.html* to obtain directions for making neurons from beads, pipe cleaners, string, or rope.

Resources for Further Exploration

Websites

Brain and Addiction
http://teens.drugabuse.gov/facts/facts_brain1.asp

Brain Function and Deficits
www.neuroskills.com/brain-injury/brain-function.php

How Your Brain Works
http://science.howstuffworks.com/brain.htm/

Nerve Function and Drug Action
www.utexas.edu/research/asrec/neuroncartoon.html

Neuroscience for Kids
http://faculty.washington.edu/chudler/neurok.html

The Inside Story of Cell Communication
http://learn.genetics.utah.edu/content/begin/cells/insidestory/

Books

Sylwester, R. 1995. *A celebration of neurons: An educator's guide to the human brain.* Alexandria, VA: ASCD.

HOW DOES THE BRAIN WORK?
TRANSPARENCY G: FUNCTIONS OF THE BRAIN

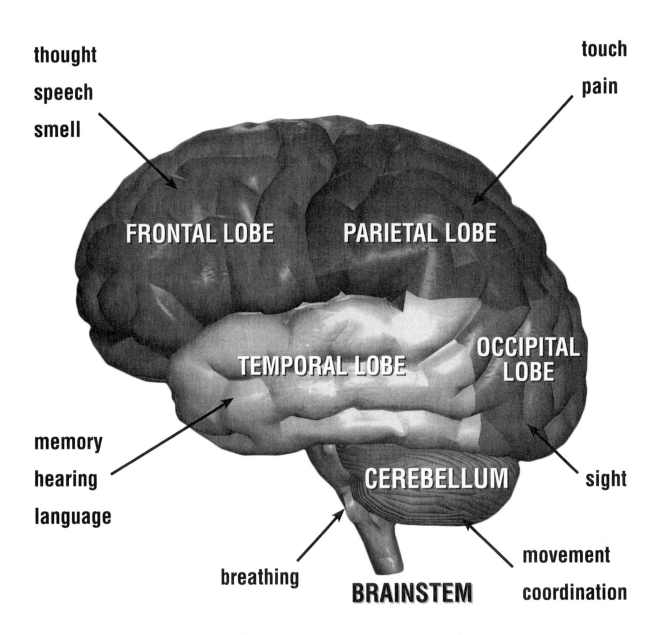

thought

speech

smell

touch

pain

FRONTAL LOBE

PARIETAL LOBE

OCCIPITAL LOBE

TEMPORAL LOBE

memory

hearing

language

CEREBELLUM

sight

breathing

BRAINSTEM

movement

coordination

HOW DOES THE BRAIN WORK?

TRANSPARENCY H: HOW A NEURON SENDS A MESSAGE

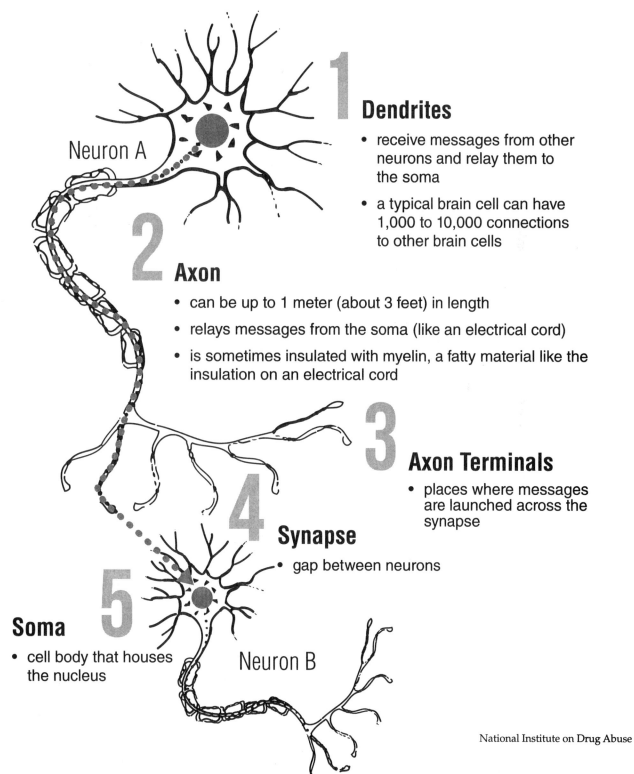

1 Dendrites

- receive messages from other neurons and relay them to the soma

- a typical brain cell can have 1,000 to 10,000 connections to other brain cells

2 Axon

- can be up to 1 meter (about 3 feet) in length

- relays messages from the soma (like an electrical cord)

- is sometimes insulated with myelin, a fatty material like the insulation on an electrical cord

3 Axon Terminals

- places where messages are launched across the synapse

4 Synapse

- gap between neurons

5 Soma

- cell body that houses the nucleus

Neuron A

Neuron B

National Institute on Drug Abuse

HOW DOES THE BRAIN WORK?
TRANSPARENCY I: SENDING THE MESSAGE ACROSS THE SYNAPSE

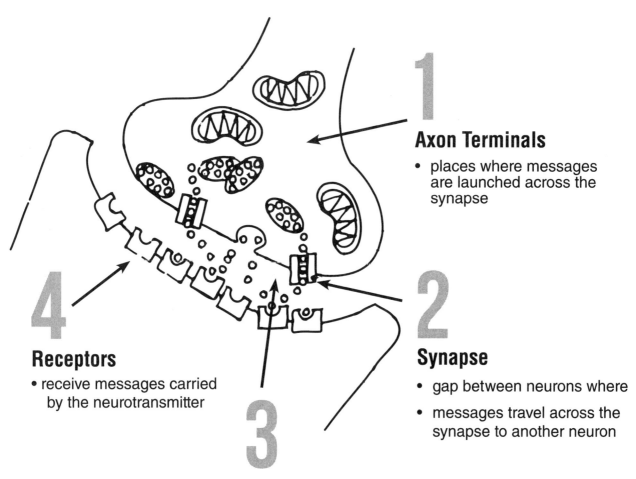

1

Axon Terminals

- places where messages are launched across the synapse

2

Synapse

- gap between neurons where
- messages travel across the synapse to another neuron

3

Neurotransmitters

- carry messages from one neuron across the synapse to another neuron

4

Receptors

- receive messages carried by the neurotransmitter

National Institute on Drug Abuse

Name_____ Date_____

FUNCTIONS OF THE BRAIN
WORKSHEET 5

_____

_____   _____

_____   _____

FRONTAL LOBE   **PARIETAL LOBE**

OCCIPITAL LOBE

TEMPORAL LOBE

CEREBELLUM

_____   _____

_____

_____   **BRAINSTEM**   _____

_____

Name_____ Date_____

HOW THE NEURON SENDS A MESSAGE
WORKSHEET 6

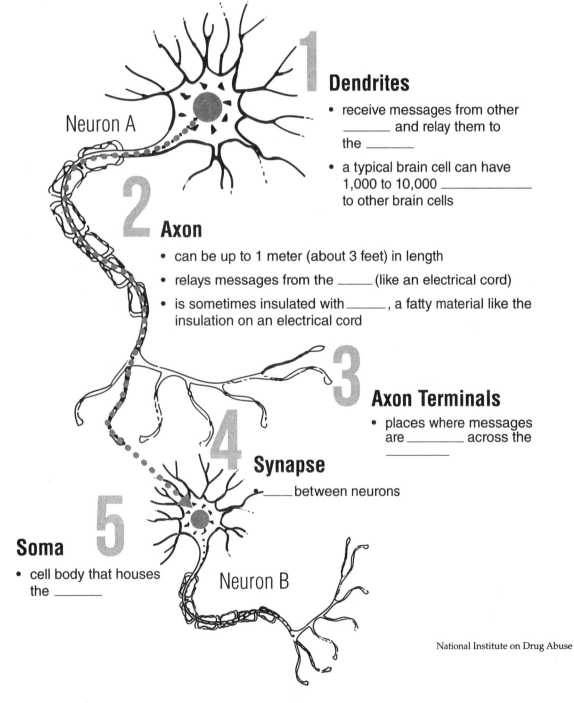

Neuron A

1 Dendrites
- receive messages from other _____ and relay them to the _____
- a typical brain cell can have 1,000 to 10,000 _____ to other brain cells

2 Axon
- can be up to 1 meter (about 3 feet) in length
- relays messages from the _____ (like an electrical cord)
- is sometimes insulated with _____, a fatty material like the insulation on an electrical cord

3 Axon Terminals
- places where messages are _____ across the _____

4 Synapse
- ____between neurons

5 Soma
- cell body that houses the _____

Neuron B

National Institute on Drug Abuse

Name_____   Date_____

CREATE A NEURON
WORKSHEET 7

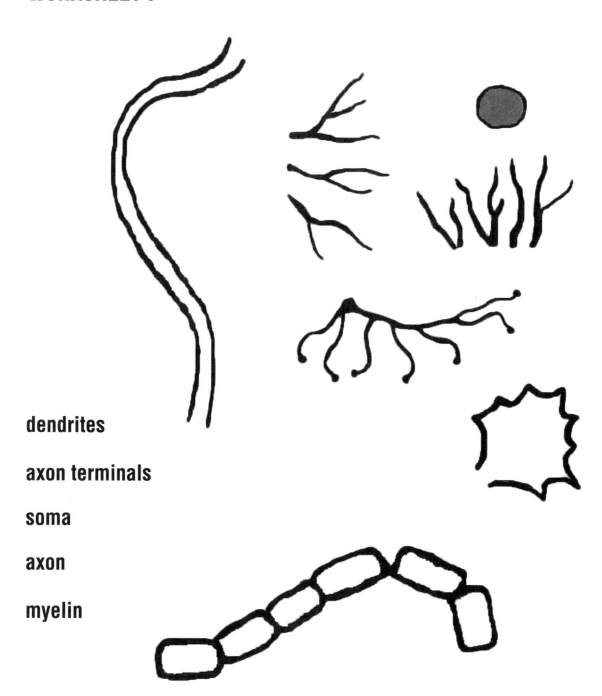

dendrites

axon terminals

soma

axon

myelin

SENDING THE MESSAGE ACROSS THE SYNAPSE
WORKSHEET 8

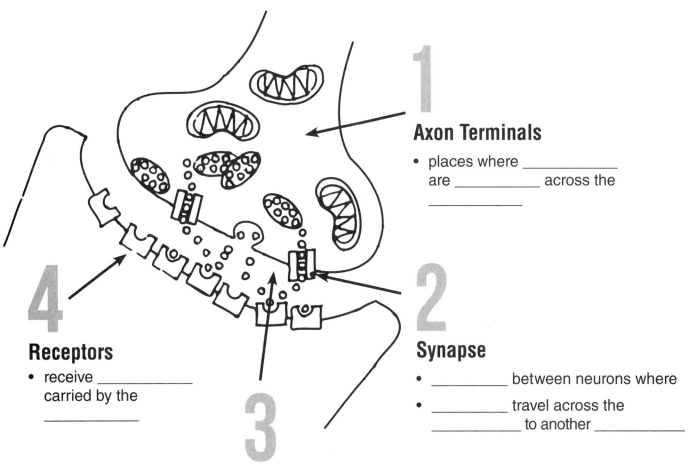

1

Axon Terminals
- places where _____ are _____ across the _____

2

Synapse
- _____ between neurons where
- _____ travel across the _____ to another _____

3

Neurotransmitters
- carry _____ from one _____ across the synapse to another _____

4

Receptors
- receive _____ carried by the _____

National Institute on Drug Abuse

Name_____ Date_____

THE BRAIN GAME CROSSWORD PUZZLE
HOMEWORK EXTENSION

Use your Lesson 3 Worksheets to help solve the puzzle.

THE BRAIN GAME CROSSWORD PUZZLE
HOMEWORK EXTENSION (CONTINUED)

ACROSS

1. Receive and relay message to the brain.

3. You could not take a breath without this area of the brain.

4. A typical brain can have thousands of these between brain cells.

7. The ability to perform jumping jacks relies on this section of the brain.

9. This gap is not a clothing store.

11. The frontal lobe helps with this important job.

12. Messages are launched across the synapse from this location.

14. The nucleus is housed in this cell body.

DOWN

2. This lobe makes hearing possible.

5. These carry messages across the synaptic gap.

6. Receives massages carried by neurotransmitters.

8. The axon is sometimes insulated with this material.

10. One of the feelings sensed by the parietal lobe.

13. Messages from the soma are relayed through this part of the neuron.

PUTTING IT ALL TOGETHER

Based on the information from this crossword puzzle and the Lesson 3 classroom activities on the brain and neurons, summarize one new insight you have that might relate to Chris's condition.

_____

_____

_____

_____

LESSON 4

HOW DO DRUGS AFFECT THE BRAIN?

Teacher Lesson Plan: Overview and Concepts

In this lesson, students build on Lesson 3's exploration of neuron structure. In addition to understanding how neurons transfer messages, students learn how drug use interferes with the normal functioning of the brain and has negative effects. Students continue to follow the story of Chris through the process of the scientific method as they gather important information about brain function and drug use. Students will also be challenged to develop an explanation of why drug addiction is considered a disease of the brain.

Students discover the process of message transfer in the brain and what happens to that process when drugs of abuse are introduced into the body. Using critical-thinking skills, students analyze the risks associated with drug use and abuse, the science of addiction, and how drug use changes the normal functioning of human body systems. This knowledge will be used in the next lesson to explore the role of scientific research in understanding drug addiction.

Objectives

After completing this lesson, students will be able to

- demonstrate the process of message transfer between specialized neuron cells;
- explain the role of neurotransmitters in message transmission;
- illustrate how drugs of abuse interfere with the communication process between neurons;
- define *disease* and formulate an explanation of why drug addiction is considered a brain disease; and
- explain the difference between drug abuse and addiction.

4 HOW DO DRUGS AFFECT THE BRAIN?

National Standards Met in Lesson 4

National Science Education Standards

Standard A: Science as Inquiry

- Abilities necessary to do scientific inquiry
- Understanding about scientific inquiry

Standard B: Physical Science

- Transfer of energy

Standard C: Life Science

- Structure and function in living systems
- Regulation and behavior

Standard F: Science in Personal and Social Perspectives

- Personal health
- Risks and benefits
- Science and technology in society

National Health Education Standards

- Students explain the relationship between positive health behaviors and prevention of injury, illness, disease, and premature death. (standard 1)
- Students explain how health is influenced by the interaction of body systems. (standard 1)
- Students predict how decisions regarding health behavior have consequences for self and others. (standard 6)

National Council of Teachers of English

- Students participate as knowledgeable, reflective, creative, and critical members of a variety of literacy communities. (standard 11)
- Students use spoken, written, and visual language to accomplish their own purposes (e.g., for learning, enjoyment, persuasion, and the exchange of information). (standard 12)

ACTIVITY ONE
BRIDGING THE GAP

Time needed for completion:

20 minutes

MATERIALS

For the class:

- Overhead projector
- Transparency J
- Small colored objects
- Colored pencils or crayons

For each student:

- Worksheet 9
- Small identical objects in two colors

Preparation

Gather 5–10 small, identical objects of one color and one of the same object in another color (e. g., buttons, beads, geometric-shaped math counters) to demonstrate normal and interrupted neural communication.

Procedure

1. Show students Transparency J: *How Have Drugs Affected Chris's Brain?* from Lesson 3. Have students share what they already know about drug use and its impact on brain function. Ask students if they think Chris's brain or body functions have been affected by drug use. Ask for specific examples to describe possible symptoms or effects Chris might be experiencing due to drug use. Examples could be high blood pressure, red eyes, sleeplessness, irritability, seizures, distractibility, poor energy, paranoia, irregular heart rhythms, difficulty in school, memory loss, or problems getting along with family and friends.

2. Explain to students that they will be using the information from Lesson 3 to discover what happens to normal brain function when drugs are introduced into the body.

3. Choose five student volunteers to create a human neuron chain. Each person will represent a neuron: the left hand represents the dendrites, the body is the soma, the right arm is the axon, and the right hand represents the axon terminal. Have students stand in a line with arms outstretched with fingertips close but not touching.

4. To initiate the transmission, place a small object of the common color in the left hand of the first person in line. That person will transfer the object to his or her right hand (signifying the signal moving through the "neuron"), and then place the object in the

hand of the person to the right. The next person transfers the object to his or her right hand and places it in the left hand of the person to the right. This sequence continues to the end of the chain. The person at the end of the chain should be instructed to perform an action signifying the successful communication (e.g., a wink, a laugh, saying a word). Discuss with the class each step of the process stressing the fit between neurotransmitter and receptor and that the receptor must be available for each neurotransmitter in order for the signal to continue down the chain.

5. To demonstrate an interrupted signal, start the human neuron chain again, this time inserting an object of the second color into the left hand of the person in the middle of the chain. Begin the process to demonstrate how the message (object coming down the chain) cannot be accepted by this neuron because there is already an object being used by the receptor.

6. Have the students get into their small groups. To review the concepts from Lesson 3, distribute this lesson's Worksheet 9: *Brain Function and Drug Use* (p. 85).

7. Have students complete only Worksheet 9: Part A: Bridging the Gap. (Encourage the students to review Worksheet 6: *How a Neuron Sends a Message* and Worksheet 8: *Sending the Message Across the Synapse* from Lesson 3 in their project portfolio pages.) In Worksheet 9: Part A (question 1), students are asked to make a detailed color drawing of a two-neuron chain, including neurotransmitters, showing how a message is passed between neurons. Students should label the dendrites, soma, axon, axon terminals, synaptic gap, neurotransmitters, and receptors, and show the direction of the electrical impulse. On their drawing, students should show how the shape of the neurotransmitter fits the shape of the receptor.

8. In the small groups, have students discuss and complete Worksheet 9: Part A (question 2).

ACTIVITY TWO
BREAKING THE CHAIN

Time needed for completion:

25 minutes

MATERIALS

For each student:

- Worksheet 9
- Internet access

Procedure

1. Ask students to *hypothesize* what may happen to message transmission in the brain if drugs of abuse are present. In their small groups, have students brainstorm a list of possible ways a drug of abuse (such as cocaine, marijuana, inhalants, alcohol, and so on) could interfere with message transmission between neurons. Encourage students to use both color objects and their drawing from Worksheet 9: *Brain Function and Drug Use,* Part A as a model. After students have had several minutes to brainstorm, have the groups share their lists with the class. Record the complete list on the board.

2. Consider displaying Transparency I: *Sending the Message Across the Synapse.* Using the list brainstormed by the class, lead the students in a discussion about how drugs can interfere with message transmission between neurons in the following basic ways:

   - The drug of abuse (such as PCP) may have a *similar size and shape* as the natural neurotransmitter and take its place on the receptor of the dendrite. If the natural neurotransmitter (frequently dopamine) cannot find its receptor because the drug has taken its place, the message cannot be delivered and the neurotransmitter builds up in the synaptic gap.

   - Some drugs (such as cocaine) can *block the reabsorption* of the natural neurotransmitter back into the axon terminal of the sending neuron, causing a flood of the neurotransmitter in the synapse.

   - Some drugs (such as methamphetamine) *bind into the receptors* and start an unusual chain of electrical impulses. This causes the neuron to release large amounts of the natural neurotransmitter into the synapse.

3. Direct students to: *www.thirteen.org/ closetohome/science/html/animations.html* to view an online animation of these effects on nerve cell transmission.

4. Explain to students that when there is an unnatural amount of a neurotransmitter present in the synaptic gap, it causes the person to have significantly increased feelings of pleasure and reward. The brain may react to the excess dopamine in several ways:

   - The brain may reduce the production of natural dopamine to compensate for the excess released due to drug use.

   - The brain may decrease the number of dopamine receptors to compensate.

   - Some neurons in the brain may die from the toxicity of the drug taken.

5. Over time, excess neurotransmitter introduced through drug abuse can damage the ability to feel pleasure and the nerve cells can become dependent on the drug to function properly.

6. Have students complete question 4 of Worksheet 9, *parts A and B* in their small groups. Have students select one color to represent the natural neurotransmitter and a different color for the drug of abuse.

7. Ask students how this process can lead to a person becoming addicted to a drug. Write the terms *abuse* and *addiction* on the board and ask students to describe the difference between their meanings.

   - Drug *abuse* is the *repeated* use of illegal drugs or the *inappropriate* use of legal drugs to increase the feelings of pleasure, reduce stress, or alter reality.

   - Emphasize to students that an *addiction* is apparent when a person has a pattern of behavior in which drugs are used and sought in an uncontrolled and compulsive way—even in the face of negative health and social consequences. When a person becomes addicted to a drug, the drug craving and seeking becomes central to the person's life. Tell students that no one can know how quickly someone can become addicted to a drug, but scientific evidence has shown that exposure to drugs in young people has a more potent addictive impact than exposures later in life. This is especially important for children to understand.

8. Ask students, "Why can it be difficult for a person to stop using a drug?" Explain that through research, scientists have learned that repeated drug use fundamentally

changes the brain so that the limbic system craves the drug just as it does food and water. That is why a person may experience withdrawal symptoms when the individual stops taking the drug. When drug use is no longer voluntary, a person has become truly addicted or "dependent" on drugs.

9. Introduce the term *disease*, which is an abnormal condition that impairs the normal functioning of an organism. Using this definition and the concepts discussed, have students analyze why drug addiction can be considered a disease of the brain. (*Note:* There is some controversy in the public regarding the "disease" concept of drug addiction. However, it is widely accepted in the biomedical and science community that drug addiction becomes a chronic health condition like asthma or heart disease with biological, psychological, and social influences. Discuss with students why there are differing views on this concept.)

10. Ask students, "If drug addiction is a disease, is there a cure?" Explain there is no cure for drug addiction, but it is treatable through programs that may use behavior therapy and sometimes medications. The purpose of treatment is to restore a healthier life by regaining control over the drug craving, seeking, and use. Significant research is being done by organizations such as the National Institute on Drug Abuse to help scientists understand the effects of drug use on the human body, how addiction occurs, and how it can be more effectively treated.

11. Have students complete the remaining items in Worksheet 9: Part B.

ACTIVITY THREE

CLOSING TEASER— TOMORROW'S HEADLINE

Time needed for completion:

5 minutes

For the class:

- Overhead projector
- Transparency K

For each student:

- Lesson 4 homework extension
- Worksheet 9

Procedure

After completing the activities, assign the homework extension, then display Transparency K: *Can Research Help Chris?* without providing additional details.

Homework Extension: The Brain Scrambler

Hand out Homework Extension: *The Brain Scrambler* (p. 88) for students to complete before the next lesson. Have students complete Worksheet 9 if they were unable to finish during class.

Optional Extensions

- Have students construct detailed 3-D models of a chain of neurons to illustrate the process of message transfer. The models should be labeled and include arrows to indicate the direction of the neural impulse.

- Assign groups of students to prepare an oral or written research report on the specific effects various drugs of abuse have on the structure and function of human body systems and particularly the brain. Selected drugs could include cocaine, marijuana, LSD, ecstasy, heroin, steroids, or methamphetamine.

- Have students create a commercial or skit illustrating the risks associated with drug use, abuse, and dependence. Students should include a scientific explanation of drug addiction and why addiction is a brain disease with biological, psychological, physiological, and social detriments. Students should include some of the negative consequences that are due to drug use and abuse, such as memory loss, chronic respiratory problems, depression, headaches, racing heart, increased blood pressure, strokes, bleeding in the brain, paranoia, aggressiveness, and sleep disorders.

Resources for Further Exploration

Websites

Centre for Neuro Skills. Brain function and deficits: *www.neuroskills.com/index.shtml?main=/tbi/brain.shtml*

HowStuffWorks. How your brain works: *http://science.howstuffworks.com/brain.htm brain.shtml*

NIDA for teens. Introducing your brain: *http://teens.drugabuse.gov/facts/facts_brain1.asp*

NIDA for teens. Mind over matter teacher's guide: *http://teens.drugabuse.gov/mom/tg_intro.asp*

Books

Friedman, D. P., S. Rusche, and J. Biswas. *False messengers: How addictive drugs change the brain*. New York: CRC Press.

Grabish, B. R. *Drugs and your brain*. Center City, MN: Hazelden.

Papa, S., and S. Miller. *Amazing brain: Addiction*. Farmington Hills, MI: Blackbirch Press.

Sylwester, R. 1995. *A celebration of neurons: An educator's guide to the human brain*. Alexandria, VA: ASCD.

HOW DO DRUGS AFFECT THE BRAIN?

TRANSPARENCY K

BRAIN FUNCTION AND DRUG USE
WORKSHEET 9

Part A: Bridging the Gap

1. Draw a detailed diagram of a two-neuron chain, including neurotransmitters, showing how a message is sent between neurons. Label the dendrites, soma, axon, axon terminals, synaptic gap, neurotransmitters, and receptors, and draw an arrow to indicate the direction of the electrical impulse. Be sure to show how the shape of the transmitter (object) fits the shape of the receptor.

2. Using the terms *message, neurotransmitters, lock, key, axon terminal, receptor,* and *synaptic gap*, explain how a message is transferred between neurons.

_____
_____
_____
_____
_____
_____
_____
_____
_____

Name_____ Date_____

BRAIN FUNCTION AND DRUG USE
WORKSHEET 9 (CONTINUED)

Part B: Breaking the Chain

3. Draw and label a colored diagram showing how a drug such as marijuana can interrupt message transfer between neurons. Be sure to include the axon terminal, synapse, neurotransmitters, and receptors.

4. Write a paragraph to explain why drug addiction is considered a brain disease. Describe some ways in which drug use may have negatively affected Chris's brain and body.

_____
_____
_____
_____
_____
_____
_____
_____
_____
_____

BRAIN FUNCTION AND DRUG USE
WORKSHEET 9 (CONTINUED)

5. Explain three ways a drug can interfere with natural message transmission between neurons.

_____
_____
_____
_____
_____
_____
_____
_____
_____
_____

6. Do you believe scientific research can help Chris? Explain why or why not.

_____
_____
_____
_____
_____
_____
_____
_____
_____

Name_____ Date_____

THE BRAIN SCRAMBLER
HOMEWORK EXTENSION

Based on the material in your Project Portfolio and the Student Glossary (p. 251), follow the clues to unscramble each of these words and write the letters on the lines provided. Rearrange the shaded letters to solve the ultimate puzzle at the bottom of the page.

a. A short description of a person's characteristics

ILEROPF __ __ ▓ __ __ __ __

b. Neurotransmitter that produces feelings of pleasure when released by the brain's reward system

MOPIDENA __ __ __ __ __ ▓ __ __

c. Specialized branches that receive messages from other neurons and relay them to the soma

STENDDRIE __ __ __ ▓ __ __ ▓ __ __

d. The addictive drug in tobacco

ENOCNTII __ __ ▓ __ __ __ __ __

e. The repeated use of drugs to increase pleasure or reduce stress

GUDR SEAUB ▓ __ __ __ __ __ __ __ __

f. The part of the body that controls all thoughts, feelings, and other body functions.

NIRAB __ __ __ __ ▓

g. Part of the neuron that contains the nucleus

OAMS __ __ __ ▓

h. A fatty material that surrounds and insulates the axons of some neurons

LYMNIE __ __ __ __ ▓ __

ULTIMATE PUZZLE:
Uncontrollable, compulsive drug seeking and use, even in the face of negative health and social consequences.

Answer: _____

LESSON 5

HOW SCIENCE IS HELPING

Teacher Lesson Plan: Overview and Concepts

In this lesson, students explore four categories of neuroscience research—specifically, four approaches that are helping to understand drug addiction—basic biological research, population-based research, behavioral research, and genetic research.

Students analyze and sort a variety of research studies that have been conducted into these four main categories. Students see how research has played an important role in understanding substance use, abuse, and addiction. To engage students in discussion, facts about each research study are presented. Students use these facts to develop an understanding of how the knowledge gained from scientific research is used in clinical practice to help people with addiction. Students begin to understand how a research project is designed and how the results can affect all of us.

Objectives

After completing this lesson, students will be able to

- work cooperatively in small groups to identify, analyze, and categorize information;

- develop an understanding of the four different approaches to scientific research;

- use research data to organize, interpret, and draw conclusions;

- develop thoughtful questions that reflect the scientific inquiry process;

- understand how scientific research influences societal changes; and

- analyze the benefits of scientific research to society.

HOW SCIENCE IS HELPING

National Standards Met In Lesson 5

National Science Education Standards

Standard A: Science as Inquiry
- Abilities necessary to do scientific inquiry
- Understanding about scientific inquiry

Standard C: Life Science
- Structure and function in living systems
- Regulation and behavior

Standard F: Science in Personal and Social Perspectives
- Personal health
- Risks and benefits
- Science and technology in society

Standard G: History and Nature of Science
- Nature of science

National Council of Teachers of English

- Students apply a wide range of strategies to comprehend, interpret, evaluate, and appreciate texts. They draw on their prior experience, their interactions with other readers and writers, their knowledge of word meaning and of other texts, their word identification strategies, and their understanding of textual features (e.g., sound-letter correspondence, sentence structure, context, graphics). (standard 3)

National Health Education Standards

- Students explain the relationship between positive health behaviors and the prevention of injury, illness, disease, and premature death. (standard 1)
- Students explain how health is influenced by the interaction of body systems. (standard 1)
- Students describe ways to reduce risks related to adolescent health problems. (standard 1)
- Students explain how appropriate healthcare can prevent premature death and disability. (standard 1)
- Students describe how lifestyle, pathogens, family history and other risk factors are related to the cause or prevention of disease and other health problems. (standard 1)
- Students analyze the validity of health information, products, and services. (standard 2)

National Council for the Social Studies

- Social studies programs should include experiences that provide for the study of people, places, and environments. (standard 3)

ACTIVITY ONE
RECENT RESEARCH FINDINGS

Time needed for completion:

25 minutes

MATERIALS

For the class:

- Overhead projector
- Transparency K
- Transparency L
- Research labels (1 copy/group)

For each student:

- Worksheet 10

Preparation

Make copies of *Research Labels* (p. 96; one set for each student group).

Procedure

1. Show Transparency K: *Can Research Help Chris?* from Lesson 4.

2. Ask students if they (or anyone they know) have ever:

    - been involved in a survey, such as a TV and/or radio rating?

    - been involved as a research subject in a research study or clinical trial?

    - heard about research studies being conducted at local hospitals or universities (perhaps on TV or in newspapers or magazines)?

3. Ask students to identify some types of diseases that scientists study using research. Ask, "Do you think that scientists research topics such as drug use, abuse, and addiction?" or "Why do scientists do research?" or "Does research have any benefit for individuals or society?"

4. Review the scientific method with the students. Have students refer to the bookmark made in Lesson 2 or the scientific method worksheets in their portfolios. Discuss how scientists state a problem or question, collect relevant information, form a hypothesis (the best answer to the question being asked), test their hypothesis through the process of research, record their results, and draw a conclusion based on the results of the research. Review with students how they have been going through the steps of the scientific method with Chris. Ask, "What hypothesis might the doctor have about Chris?" or "Can research help Chris?"

5. Give each student a copy of Worksheet 10: *Research Categories* (p. 100). Explain that there are four main types of scientific research that are helping us understand various health issues: basic biological research, population-based research, behavioral research, and genetic research. Definitions for each of these categories are provided for the students on their worksheets. Have students read each definition out loud to the class. Check for understanding by asking students to provide an example for each category and making sure that students understand all of the terms used in each definition.

6. Tell students that they will work in their groups to play a game. You will reveal to the class 18 questions that have been or could be used to guide scientific research studies. The groups will have up to 30 seconds to determine whether the research is biological, population-based, behavioral, or genetic in nature. When time has expired, have each group hold up their answer (research label). Groups will get 10 points for each correct answer and 5 points will be deducted for each incorrect answer. Check for student understanding after each fact by asking, "Why was this study categorized as (include particular type) research?" You may want to decide to provide an incentive to your students.

7. Start the game by displaying the first research question on Transparency L: *Research Questions* on the overhead projector.

ACTIVITY TWO
RESEARCH AND SOCIETY

Time needed for completion:

15 minutes

MATERIALS

For each student:

• Worksheet 11

Procedure

How does the knowledge gained from scientific research get used to help people? Do *you* benefit from research? In this activity, students will look at one research study and discuss its benefits to society.

1. Give each student a copy of Worksheet 11: *Research and Marijuana Use* (p. 101). It may be helpful to make a transparency of this page.

2. As a class, discuss and identify the subjects of study (*who*). Guide the class toward the conclusion that they need two subject groups. One group, marijuana users, comprises the *research subjects* or volunteers. A second group, non-marijuana users, comprises the *control group*. Also, stress the importance of creating specific definitions for your study (e.g., How will you define *marijuana users*? What age groups might be best to look at? Why?)

3. As a class, discuss and identify the category of research that would be used to answer the study questions (*what*). The category should be identified as *Biological Research*.

4. Discuss some possible research questions. Students should list questions about topics, such as the long-term effects of marijuana use, types of cancer associated with tobacco or marijuana use, or how quickly breathing problems occur when using marijuana.

5. Discuss *how* this research has benefited society. Answers may vary, but can include the Surgeon General's warning on products containing tobacco, advancements in treatments for tobacco or marijuana-related diseases, and improved health education for children and adults.

Note: Make sure students understand Worksheet 11 before passing out the homework extension (see next page).

ACTIVITY THREE

CLOSING TEASER— TOMORROW'S HEADLINE

Time needed for completion:

5 minutes

MATERIALS

For the class:

- Overhead projector
- Transparency M

For each student:

- Lesson 5 homework extension

Procedure

After completing the activities, assign the homework extension, then display Transparency M: *Can a Mouse Help Chris?* without providing additional details.

Homework Extension: You Make the Call

Hand out Homework Extension: *You Make the Call* (p. 102) to complete before the next lesson. In this activity, students will look at a research question, much like they did in Worksheet 11: Activity 2. Students will be responsible for designing a study that could answer the research question.

Optional Extensions

- Ask students if research studies have affected their lives. Have students write a short impromptu paragraph in class on, "What research has done for me."

- Have students do research and then write a one- to two-page essay that includes actual examples of how applied research has affected their lives (e. g., Surgeon General's warnings, instructions on over-the-counter medicines, disclaimers in ads, safety of everyday household products).

- Have students do research by calling (or visiting the website of) local hospitals and universities to learn about what kinds of medical research and/or clinical trials are being conducted at local institutions.

- Have students work in their small groups or independently to come up with one question for each category of research (biological, genetic, population-based, and behavioral) about Chris. The class could write all of the questions down on the board/large piece of paper or in their Project Portfolios to be referred to throughout the rest of the unit. (Groups may refer back to their research questions when they explore the ethics lessons and develop possible ethical problems that would have to be addressed if they were to study their research questions.)

Resources for Further Exploration

Infofacts: The National Institute on Drug Abuse has created InfoFacts sheets on common drugs such as tobacco, alcohol, marijuana, cocaine. These can be found in the NIDA InfoFacts section of the NIDA website.

www.drugabuse.gov/publications/term/160/ InfoFacts

NIDA Notes: As stated on the National Institute on Drug Abuse website, "NIDA Notes covers drug abuse research in the areas of treatment and prevention, epidemiology, neuroscience, behavioral science, health services, and AIDS. The publication reports on research; identifies resources; and promotes communication among clinicians, researchers, administrators, policymakers, and the public."

www.nida.nih.gov/NIDA_Notes/NNIndex.html

Substance Use: This is the substance abuse section of the RAND Corporation's website. RAND Corporation is a nonprofit research organization that conducts research to address issues that affect the world.

www.rand.org/research_areas/substance_abuse

Hopkins Medicine Today: The John Hopkins Bayview Medical Center website provides links to current articles about topics that are new in healthcare.

www.hopkinsmedicine.org/news

RESEARCH LABELS

One copy for each student group. Cut on lines to make separate labels.

Biological

Behavioral

Genetic

Population-Based

HOW SCIENCE IS HELPING
TRANSPARENCY L

Research Questions

Study 1: How are neurotransmitters in the brain affected by cocaine addiction?

Study 2: How many eighth-grade students in the United States have tried alcohol at least once?

Study 3: Are there differences between the brains of people who take risks and people do not take risks?

Study 4: Are children of alcohol-dependent parents at greater risk for becoming alcohol-dependent than children of nonalcohol-dependent parents?

Study 5: How does drug dependence relate to problems in how the brain's neurons transmit messages?

Study 6: How can using different rewards/consequences reduce cigarette use?

Study 7: How do specific neuron pathways reinforce the effects of drugs?

Study 8: How can studies of twins help us look for inherited risk factors for substance abuse?

Study 9: How do heroin users get the drug into their systems?

HOW SCIENCE IS HELPING

TRANSPARENCY L (CONTINUED)

Research Questions

Study 10: Do kids try marijuana just as often as they do cigarettes?

Study 11: Are middle school students more likely to grow up to be smokers if one or both of their parents smoke?

Study 12: If a pregnant woman drinks alcohol, will her baby's development be affected?

Study 13: Is marijuana addictive?

Study 14: Are middle and high school students who use smokeless tobacco more likely to become cigarette smokers than students who don't smoke at all?

Study 15: How does genetics play a role in substance abuse and its risk factors?

Study 16: How many high school seniors become regular substance users?

Study 17: How are adolescents influenced by other adolescents and adults who smoke?

Study 18: Do regular marijuana users develop breathing problems?

HOW SCIENCE IS HELPING
TRANSPARENCY M

RESEARCH CATEGORIES
WORKSHEET 10

Activity 1

Basic Biological Research: (Neurobiological)

- Biological research involves the study of the human body and all of its systems and functions.

- Neuroscience research involves the study of the brain and the nervous system (i.e., spinal cord, nerves, nerve circuits).

Behavioral Research:

Behavioral research is the scientific study of behavior. It measures observed behaviors, along with the role of environmental factors. Behavioral research helps us understand and predict behavior. Studies have shown that behaviors are influenced by their consequences. For example, students might choose a better behavior depending on the punishment or consequence (e.g., trip to principal's office, writing a sentence 100 times).

Genetic Research:

Genetic research is the scientific study of heredity. Parents pass on genetic information to their biological children through DNA. DNA is organized into genes and then chromosomes.

Population-Based Research:

Population-based research studies a representative sample of a target population. The research results are then analyzed and applied to the entire target population.

RESEARCH AND MARIJUANA USE
WORKSHEET 11

Activity 2

QUESTION: Do regular marijuana users develop breathing problems? (Study 18)

Fact: Regular marijuana users develop breathing problems including coughing and wheezing, and are more vulnerable to lung infections. Marijuana contains the same cancer-causing chemicals as tobacco. The amount of tar and carbon monoxide inhaled by marijuana smokers is three to five times greater than the levels of tobacco smokers.

WHO
Who were the subjects of study? *Hint:* There are two groups.

_____

WHAT
What did the researchers study? _____

Research Category: _____

What are some research questions you would ask? _____

_____

HOW
How could the researcher apply this information to benefit society?

_____

Name_____   Date_____

YOU MAKE THE CALL
HOMEWORK EXTENSION

Directions: Use Worksheets 10 and 11 as a guide to complete the following:

QUESTION: Are athletes more likely than nonathletes to abuse drugs?

WHO
Who would be the subjects of study? *Hint:* There are two groups.

_____

WHAT
What would YOU study?

Research Category: _____

What data would you collect from each group? _____

HOW
How might this information benefit other people?

_____

_____

Based on the information you learned in this lesson, which category of research do you think would help Chris? Explain your answer.

_____

_____

LESSON 6
WHY RESEARCH IS IMPORTANT

Teacher Lesson Plan: Overview and Concepts

In this lesson, students work in small groups to analyze and sort a variety of research discoveries. This unit illustrates in part the historical role animals have played in medical research. Students sort these discoveries into several categories, including time period and type of animal included in the research. Students see how animals have allowed scientists and doctors to better understand many complex medical problems, which can lead to treatments and cures that help *both* animals and humans. Through animal models, researchers are able to test hypotheses to advance our understandings of disease and treatment. Animal models have contributed greatly to medical progress for many diseases such as cancer, diabetes, HIV-related diseases, and drug and alcohol addiction, Alzheimer's disease, and Parkinson's disease. In this lesson, students work in their groups to answer questions about animals in medical research and to consider which animals might help Chris. Finally, students examine how and why specific animals are selected for research based upon the animal/ organism's biological complexity and similarity to humans.

Objectives

After completing this lesson, students will be able to

- work cooperatively in small groups to identify, analyze, and categorize information;
- develop an understanding of the historical contributions of animals in medical research;
- use research data to organize, interpret, and draw conclusions;
- develop thoughtful questions that reflect the scientific inquiry process;
- understand how scientific research influences societal changes; and
- analyze the benefits of scientific research to society.

6 WHY RESEARCH IS IMPORTANT

National Standards Met in Lesson 6

National Science Education Standards

Standard A: Science as Inquiry

- Abilities necessary to do scientific inquiry
- Understanding about scientific inquiry

Standard C: Life Science

- Structure and function in living systems
- Regulation and behavior

Standard F: Science in Personal and Social Perspectives

- Personal health
- Risks and benefits
- Science and technology in society

Standard G: History and Nature of Science

- Science as a human endeavor
- Nature of science

National Health Education Standards

- Students explain the relationship between positive health behaviors and the prevention of injury, illness, disease, and premature death. (Standard 1)
- Students analyze how environment and personal health are interrelated. (Standard 1)

National Council of Teachers of Mathematics

- Students formulate questions that can be addressed with data and collect, organize, and display relevant data to answer questions.
- Students develop and evaluate inferences and predictions that are based on data.

National Council of Teachers of English

- Students employ a wide range of strategies as they write and use different writing process elements appropriately to communicate with different audiences for a variety of purposes. (standard 5)
- Students use a variety of technological and information resources to gather and synthesize information and to create and communicate knowledge. (standard 8)

ACTIVITY ONE

ANIMALS AND MEDICAL DISCOVERIES

Time needed for completion:

30 minutes

MATERIALS

For the class:

- Overhead projector
- Transparency M
- Five large sheets of poster paper
- Animal cards (1 sheet per group)

For each student:

- Worksheet 12
- Scissors
- Tape

Preparation

Before the lesson begins, place five large sheets of poster paper around the room. Each sheet should be labeled with a century starting from the 1600s to 2000.

Procedure

In this activity, students will learn about some of the ways animals have helped us understand complex medical problems. Students will sort through data to complete animal cards that list studies and their discoveries. They will then categorize their study according to the century in which it was completed.

1. Display Transparency M: *Can a Mouse Help Chris?* from Lesson 5.

2. Ask the students if they know of any specific examples of how animals have made contributions to medical discoveries. Examples include the development of vaccines and antibiotics to treat meningitis, chicken pox, and pneumonia and development of treatments for kidney failure, diabetes, hypertension, cancer, and other diseases. If students want to begin discussing the ethical issues surrounding animal inclusion in medical research, encourage the students to write down their thoughts and questions for further discussion in activity 2 of this lesson as well as for the remaining lessons of the unit.

3. Have the students go into their regular small groups. Give each group Worksheet 12: *Animals and Medical Discoveries* (p. 116).

4. Assign each group one of the following animal/groups to research and pass out one sheet of the corresponding animal cards per group:

    - rats, mice, etc.
    - cows, sheep, and pigs
    - frogs, fish, reptiles, and birds
    - rabbits

5. Let the students know that they will be responsible for looking through the data contained in Worksheet 12: *Animals and Medical Discoveries* to identify studies in which their animal group helped scientists make medical discoveries. Once students have identified these studies, they should use this information to complete the appropriate animal cards. They will have approximately 10 minutes to complete this part of this activity. See below for an example:

6. Next, students should place their animal cards on the appropriate century poster and work in their groups to answer the questions on Worksheet 12: *Animals and Medical Discoveries*.

Rodents *(rats, mice, etc.)*

Date of Study: 1954
Medical Discovery: Scientists cultured poliovirus. This led to the development of a vaccine for polio.

ACTIVITY TWO
WHAT'S YOUR OPINION?

Time needed for completion:

10 minutes

MATERIALS

For each student:

- A copy of the Student Glossary (pp. 212–221)

In this activity, students will begin to explore some basic ethical issues surrounding the inclusion of animals and humans in research.

Procedure

1. Have students use the Student Glossary to define the term *model*. Ask students to use this definition to explain what an animal model is in research and why a scientist might use an animal model for a study.

2. Explain to students that there are numerous questions, concerns, and limitations that arise when including *either* humans or animals in research and scientists must consider the pros and cons of all positions when conducting a study.

3. Using a T-chart, brainstorm a list of pros and cons of conducting research including animals *and* humans on the board. Ask students to share any of the questions they wrote down in Activity 1.

Humans		Animals	
Harms	Benefits	Harms	Benefits

4. To further assess the students' background knowledge and opinions, the following additional questions may be posed:

   - How does a scientist decide the procedure he or she will follow when conducting an experiment and if an animal or human is even necessary to the study?

   - How are humans and animals protected in scientific studies?

   - Can an animal model always be used in place of a human model or are there instances when a human might be preferred over an animal model? Are there times when neither an animal nor a human should be used in a study?

   - Why do some people object to including animals in research? How is this different from other uses of animals in our society?

   - Should or can a scientist do any experiment he or she wants that includes an animal or a human?

5. Explain that these questions are about the concept of *ethics*, which will be explored in the remaining lessons of the unit. Clarify that there are some studies that cannot be conducted using the following:

   - *Human research* is often not appropriate, particularly when use of illegal drugs is involved.

   - *Alternative research methods,* such as computer simulations, MRIs, mathematical modeling, genetic markers, and more (which are discussed further in Lesson 10), are not always effective.

6. When animal research is viewed as a potential option, there are many questions that must be systematically and thoughtfully answered and specific guidelines followed before it can even be considered. Stress to students that no one wants to subject animals to experimentation and possible distress, pain, or death unless it is absolutely necessary to accomplish an important goal. Lesson 9 contains important information about animal protections.

ACTIVITY THREE

CLOSING TEASER— TOMORROW'S HEADLINE

Time needed for completion:

5 minutes

MATERIALS

For the class:

- Overhead projector
- Transparency N

For each student:

- Lesson 6 homework extension
- Internet access

Procedure

After completing the activities, display Transparency N: *Doctor Questions Chris's Best Friend* without providing additional details.

Homework Extension: What Do You Think?

Hand out Homework Extension: *What Do You Think?* for students to complete before the next lesson. Students will need internet access.

Optional Extensions

- Have students read the following articles in the February 1997 issue of *Scientific American*: "The Benefits and Ethics of Animal Research," "Animal Research is Wasteful and Misleading," "Animal Research is Vital to Medicine," and "Trends in Animal Research." Have students write a newspaper article or editorial that either supports or argues against continuing animal research.

- Have students use the article "Evolution of Animal Use in Research" (also from the February 1997 issue of *Scientific American*) to create a timeline listing occurrences that have affected the inclusion of animals in research. How does this timeline compare to the timeline of medical discoveries that resulted from the inclusion of animals in research?

Resources for Further Exploration

Websites

Animal Models-Part 1: Behavior and Physiology: This entire edition is dedicated to how animal models in alcohol research help scientists better understand the complex mechanisms of alcoholism.
http://pubs.niaaa.nih.gov/publications/arh24-2/toc24-2.htm

Foundation for Biomedical Research: Nobel Prizes for medical and physiological breakthroughs involving animal research are noted with reference to specific animals involved in each discovery.
www.fbresearch.org/TwoColumnWireframe.aspx?pageid=128&terms=nobel%20prizeeducation/nobels.htm

Kids4Research: Responsible laboratory animal care standards are provided to teachers and students along with information on benefits of animal research to animals, humans, and the environment. Site focus includes biomedical/biological research and education.
www.kids4research.org

Understanding Animal Research Timeline: This timeline of animal contributions to medical treatment and technique development features breakthroughs from pre1900s through the 1990s.
www.understandinganimalresearch.org.uk/your_health/timeline

Books

Animal Models of Drug Addiction: Koob, G. F. 1995. Animal models of drug addiction, in *Psychopharmacology: The Fourth Generation of Progress.* 4th ed. San Diego, CA: Raven Press.
www.acnp.org/g4/GN401000072/Default.htm

Rodents *(rats, mice, etc.)*

Date of Study: _____
Medical Discovery:
_____
_____
_____

Rodents *(rats, mice, etc.)*

Date of Study: _____
Medical Discovery:
_____
_____
_____

Rodents *(rats, mice, etc.)*

Date of Study: _____
Medical Discovery:
_____
_____
_____

Rodents *(rats, mice, etc.)*

Date of Study: _____
Medical Discovery:
_____
_____
_____

Rodents *(rats, mice, etc.)*

Date of Study: _____
Medical Discovery:
_____
_____
_____

Rodents *(rats, mice, etc.)*

Date of Study: _____
Medical Discovery:
_____
_____
_____

6

Cows, Sheeps, and Pigs

Date of Study: _____
Medical Discovery:
_____
_____
_____

Cows, Sheeps, and Pigs

Date of Study: _____
Medical Discovery:
_____
_____
_____

Cows, Sheeps, and Pigs

Date of Study: _____
Medical Discovery:
_____
_____
_____

Cows, Sheeps, and Pigs

Date of Study: _____
Medical Discovery:
_____
_____
_____

Cows, Sheeps, and Pigs

Date of Study: _____
Medical Discovery:
_____
_____
_____

Cows, Sheeps, and Pigs

Date of Study: _____
Medical Discovery:
_____
_____
_____

Frogs, Fish, Reptiles, and Birds

Date of Study: _____
Medical Discovery:
_____
_____
_____

Frogs, Fish, Reptiles, and Birds

Date of Study: _____
Medical Discovery:
_____
_____
_____

Frogs, Fish, Reptiles, and Birds

Date of Study: _____
Medical Discovery:
_____
_____
_____

Frogs, Fish, Reptiles, and Birds

Date of Study: _____
Medical Discovery:
_____
_____
_____

Frogs, Fish, Reptiles, and Birds

Date of Study: _____
Medical Discovery:
_____
_____
_____

Frogs, Fish, Reptiles, and Birds

Date of Study: _____
Medical Discovery:
_____
_____
_____

Rabbits

Date of Study: _____
Medical Discovery:
_____
_____
_____

Rabbits

Date of Study: _____
Medical Discovery:
_____
_____
_____

Rabbits

Date of Study: _____
Medical Discovery:
_____
_____
_____

Rabbits

Date of Study: _____
Medical Discovery:
_____
_____
_____

Rabbits

Date of Study: _____
Medical Discovery:
_____
_____
_____

Rabbits

Date of Study: _____
Medical Discovery:
_____
_____
_____

Name_____ Date_____

ANIMALS AND MEDICAL DISCOVERIES
WORKSHEET 12

Directions: Use the data in the tables below to identify studies in which the animal or animal group that you have been given has helped scientists make medical discoveries. Use the information provided to complete the appropriate animal cards. Next place the animal cards on the appropriate century posters displayed around the classroom and complete the questions.

Date	Scientist(s)	Animal(s) Used	Medical Contribution/Discovery
Early 1600s		Dogs, sheep	First attempts at blood transfusion
1650s		Rodents	Need of oxygen in respiration
Late 1600s		Rabbits	First cataract surgeries
Early 1700s		Toad	Treatment for edema (swelling) studied
Mid to Late 1700s	Jenner	Cow, sheep, pigs	Development of smallpox vaccine
1810s		Dogs, sheep	Discovery of the parathyroid gland
1885	Pasteur	Dog	Development of rabies vaccine
1902	Ross	Pigeon	Understanding of malaria life cycle
1904	Pavlov	Dog	Behavioral responses to different stimuli
1912	Carrel	Dog	Advances in surgical techniques for suturing and grafting blood vessels
1923	Banting, Maclead	Dog, rabbit, fish	Discovery of insulin and mechanisms of diabetes
1936	Dale, Loewi	Cat, frog, bird, reptile	Chemical transmission of nerve impulses
1940s		Monkeys, chimpanzees, etc.	Discovery of Rhesus (Rh) factor in blood (positive or negative factor in blood type, i.e., O+ vs. O–)
1944	Erlanger, Gasser	Cat	Function of nerve cells
1945	Fleming, Chain, Florey	Mouse	Discovery of penicillin and how it can cure some infectious diseases

ANIMALS AND MEDICAL DISCOVERIES
WORKSHEET 12 (CONTINUED)

Date	Scientist(s)	Animal(s) Used	Medical Contribution/Discovery
1950s		Monkeys, chimpanzees, etc., rabbits, rodents	Therapeutic use of chemotherapy for cancer and leukemia
1954	Enders, Weller, Robbins	Monkey, mouse	Development of the yellow fever vaccine
1954		Rodents	Poliovirus cultured which led to the development of a vaccine for polio
1966	Rous, Huggins	Rat, rabbit, hen	Tumor-inducing viruses and hormonal treatment of cancer
1968	Block, Lynen	Rat	Interpretation of genetic code and its role in protein synthesis
1970	Katz, von Euler, Axelrod	Cat, rat	Mechanisms of the storage and release of nerve transmitters
1970s		Monkeys, chimpanzees, etc.	Development of rubella and hepatitis B vaccine
1970s		Dog	Discovery of the links between cholesterol and heart disease
1972	Edelman, Porter	Guinea pig, rabbit	Chemical structure of antibodies
1973	Von Frisch, Lorenz, Tinbergen	Bee, bird, fish	Organization of social and behavior patterns in animals
1986	Levi-Montalcini, Cohen	Mouse, chick, snake	Nerve growth factor

Name_____ Date_____

ANIMALS AND MEDICAL DISCOVERIES
WORKSHEET 12 (CONTINUED)

Date	Scientist(s)	Animal(s) Used	Medical Contribution/Discovery
1990s		Monkeys, chimpanzees, etc.	Advances in our understanding of AIDS
1990s		Monkeys, chimpanzees, etc.	Alleviation of neurological symptoms of Parkinson's Disease
1990	Murray, Thomas	Dog	Organ transplantation techniques
1991	Neher, Sakmann	Frog	Chemical communication between cells
1996	Prusiner	Mouse, hamster	Discovery of prions, a new biological principle of infection
2000	Carlsson, Greengard, Kandal	Sea slug, mouse	Discoveries about signal transduction in the nervous system
2003	Lauterbur, Mansfield	Clam, mouse, dog, rat, chimpanzee, pig, rabbit, frog	Discoveries about magnetic resonance imaging (MRI)

Facts from the American Association for Laboratory Animal Science and UTHSCSA Posters

ANIMALS AND MEDICAL DISCOVERIES
WORKSHEET 12 (CONTINUED)

1. **List two examples of how each animal or animal group listed below has contributed to advances in medical science:**

    Birds, Frogs, and Reptiles _____

    Cats _____

    Cows, Sheep, and Pigs _____

    Dogs _____

    Monkeys, chimpanzees, etc. _____

    Rabbits _____

    Rodents _____

2. **Which type of animal do you think is most commonly included in medical research? Why?**

    _____
    _____
    _____
    _____
    _____
    _____
    _____
    _____
    _____
    _____
    _____
    _____
    _____

Name_____ Date_____

ANIMALS AND MEDICAL DISCOVERIES
WORKSHEET 12 (CONTINUED)

3. Why do you think animals are included in medical research?

_____
_____
_____
_____
_____
_____
_____
_____
_____
_____
_____

4. Do you think animals also benefit from research? Can you think of any examples?

_____
_____
_____
_____
_____
_____
_____
_____
_____
_____
_____
_____

ANIMALS AND MEDICAL DISCOVERIES
WORKSHEET 12 (CONTINUED)

5. Can you think of ways in which animal research could be important in learning about drug abuse and addiction?

_____
_____
_____
_____
_____
_____
_____
_____
_____
_____
_____
_____
_____
_____
_____
_____
_____
_____
_____

Name_____ Date_____

WHAT DO YOU THINK?
HOMEWORK EXTENSION

Directions: Visit the website *www.kids4research.org*.
Use the information on these web pages to answer the following questions.

1. Which type of animal is most commonly included in medical research? (List the percentages.) Why do you think this animal is included the most?

    _____
    _____
    _____
    _____
    _____
    _____
    _____
    _____
    _____
    _____
    _____

2. Why have mice been included in research to study cancer?

    _____
    _____
    _____
    _____
    _____
    _____
    _____
    _____
    _____
    _____
    _____

WHAT DO YOU THINK?

HOMEWORK EXTENSION (CONTINUED)

3. **How have rabbits been included in research to study drug abuse and addiction?**

_____

_____

_____

_____

_____

_____

_____

_____

_____

_____

_____

4. **Which animals have been helpful in studying HIV/AIDS?**

_____

_____

_____

_____

_____

_____

_____

_____

_____

_____

_____

_____

Name_____ Date_____

WHAT DO YOU THINK?
HOMEWORK EXTENSION (CONTINUED)

5. How have domesticated cats and dogs benefited from animal research?

_____
_____
_____
_____
_____

6. How have other types of animals helped medical science?

Ferret_____

Chinchilla_____

Armadillo_____

Lobster _____

Opossum _____

7. Visit the website *www.mismr.org/educational/drugabuse.html*. Use the information contained within these web pages to summarize one way in which animal research will help scientists and health care professionals learn more about drug abuse and addiction.

_____
_____
_____
_____
_____
_____
_____

LESSON 7

WHAT IS ETHICS IN SCIENCE?

Teacher Lesson Plan: Overview and Concepts

In this lesson, students look at and clarify their understanding of the concept of ethics. In the first activity, students develop a definition for *ethics* based on their prior knowledge. Then as a class, students explore the meaning of five key ethics principles and apply these principles to a hypothetical situation. Working in their small groups, students discuss and evaluate a dilemma to determine what might be a sound and careful ethical response for the situation. After completing the small-group activity, students discuss and debate their group responses for the situation in a class discussion. Following this discussion, their attention will be focused back to the term *ethics* to develop a class definition.

Students should be encouraged to explore and express their ethical position regarding the situation; however, students should also be able to defend their positions and choices in a systematic manner using the ethics principles introduced in this lesson.

Objectives

After completing this lesson, students will be able to

- clarify their understanding of the term *ethics*;
- demonstrate an understanding of the following terms: autonomy, beneficence, compassion, justice, non-maleficence, veracity, and bioethicist;
- listen critically to the opinions and arguments of others;
- develop a concise and thorough definition of *ethics*;
- use analytical skills to explore and evaluate their understanding of ethical issues and principles in hypothetical situations; and
- exhibit an increased sense of personal and social responsibility.

7 WHAT IS ETHICS IN SCIENCE?

National Standards Met in Lesson 7

National Science Education Standards

Standard A: Science as Inquiry

- Abilities necessary to do scientific inquiry
- Understanding about scientific inquiry

Standard F: Science in Personal and Social Perspectives

- Personal health
- Risks and benefits

Standard G: History and Nature of Science

- Science as a human endeavor
- Nature of science

National Council for the Social Studies

- Social studies programs should include experiences that provide for the study of individual development and identity. (standard 4)

National Council of Teachers of English

- Students employ a wide range of strategies as they write and use different writing process elements appropriately to communicate with different audiences for a variety of purposes. (standard 5)

Background for Teachers

An essential component of the middle school science curriculum is to introduce students to the relationships between science and ethics. Recently developed state and national goals, objectives, and standards stress the importance of including topics that incorporate ethics/values education in the science classroom. Often, students have little to no experience learning about ethics and ethical issues placed within the context of science. Middle school students are at an important age for expanding their reasoning and logical skills, problem-solving abilities, moral values, and capacity for self-reflection.

Due to recent advances in areas such as biology, chemistry, medicine, and environmental science, society is continuously confronted with challenging ethical and social dilemmas about what is "acceptable." Carefully and systematically analyzing various situations of ethical importance is an essential skill for students to develop.

The overall purpose of the unit is for students to work through the relationship of ethics issues and scientific work in our society. Students will learn about our society's accepted guidelines for the ethical inclusion of animals in research, highlighting drug abuse research as a key example. As a result, students will be able to develop an understanding of the roles of science and animal models in developing treatments for drug abuse and misuse. In order to do so, students must first understand the concept of ethics. After introducing the key ethical principles, students begin to understand that while there is a lot of agreement among people regarding the importance of these principles, reasonable people can, and often do, come to different conclusions about what is and is not ethical.

This lesson is designed to encourage students to self-reflect, analyze situations, listen critically to other's views, and develop a comprehensive understanding of the concept of ethics. The aim is to encourage flexible approaches to evaluating ethical issues, rather than to arrive at a single "right answer." In the area of research involving animals, there are many opinions and concerns, and our society as a whole has come to the conclusion that animals may be included in research ethically only under certain very strictly defined circumstances. Assuring that these criteria are met is carefully monitored. Through this lesson, students are given the opportunity to further develop their skills related to creative problem-solving, conflict resolution, and compromise.

ACTIVITY ONE
WHAT IS ETHICS?

Time needed for completion:

10 minutes

MATERIALS

For the class:

- Overhead projector
- Transparency N

For each student:

- Worksheet 13

Procedure

1. Display Transparency N: *Doctor Questions Chris's Best Friend* from Lesson 6. Ask students why they think the doctor might want to talk to Chris's best friend. Engage the students in a discussion about the kinds of questions the doctor might ask. Questions could include, "Did you know Chris was taking drugs?" or "Have you noticed any changes in Chris's behavior lately?" or "Do you know what kind of drugs Chris was taking?" or "Did you tell anyone that Chris was using drugs?" Be sure to stress to students that doctors, in reality, *cannot* talk to your friends without your permission. This image is only meant to demonstrate that the scientific process involves gathering information. In our story about Chris, this is a way to gather more information to create an informed hypothesis about Chris.

2. Explain to students that in this lesson, the concept of *ethics* will be explored. Ask students if they believe Chris's best friend should answer the doctor's questions. Stress from the start that students should not simply state their opinion, but they must also provide evidence that supports that opinion. Have the students briefly explain their positions.

3. Distribute one copy of Worksheet 13: *What Is Ethics?* (p. 135) to each student. To determine the student's prior knowledge, have the students answer the question, "What does the word *ethics* mean to you?" independently. Allow several minutes for the students to think about and construct their own definitions. If any students are unsure, encourage them to write down their thoughts and let them know there are no wrong answers.

ACTIVITY TWO
WHAT WOULD YOU DO?

Time needed for completion:

20 minutes

MATERIALS

For the class:

- Overhead projector
- Transparency O

For each student:

- Worksheet 13
- Worksheet 14

Procedure

1. To prepare students for their small-group discussions, display Transparency O: *Key Ethics Terms*. Discuss the meaning of each term and provide an example of each. Have the students record the examples for these terms on the chart on Worksheet 13: *What Is Ethics?*

2. Read the introduction to Worksheet 14: *What Would You Do?* (p. 136) aloud and establish ground rules for the discussion to ensure a safe learning environment. (Sample group rules include: (1) Only one student talks at a time, and (2) Students use "I" statements when discussing the issues, etc.) It is essential that students understand that all ideas and thoughts are equally worthy of being expressed. The purpose of this activity is for students to use their prior knowledge and key ethics principles to generate new ideas and to explore other people's understandings regarding the concept of "ethics" in a hypothetical situation.

3. As a class, have the students read "To Tell or Not to Tell..." in Worksheet 14.

4. Have the students gather in their small groups to complete the questions. Assign each student a specific role, such as group reader, recorder, mediator, or group spokesperson.

5. Have the students record their answers in the spaces provided. This process will allow students to organize and explain their thoughts of how to ethically resolve situations. Students should be encouraged to share and compare their ideas with others in order to develop a deeper personal understanding of the concept of ethics.

6. After each group has completed the questions, discuss the group responses. Each group should be able to defend their decisions based on the key ethics principles.

ACTIVITY THREE
CLASS DEFINITION OF ETHICS

Time needed for completion:

10 minutes

MATERIALS

For each student:

- Worksheet 13

Procedure

1. For this activity, have the students brainstorm a list of words, terms, explanations, etc. of what they now believe the term *ethics* means. Write down the students' responses until all ideas have been recorded, even if you disagree.

2. Ask the class to analyze the list and discuss the recorded ideas.

3. After the class agrees on a list of applicable concepts, ask the class to construct a revised definition of *ethics* and record their definition at the top of Worksheet 13. The definition may include statements such as:

Ethics refers to ways of understanding and examining moral issues that are shaped by individual, community, and societal values. Ethical reasoning involves a systematic process that generates acceptable, justifiable choices or options.

Students should also understand that ethics requires observation, awareness, reflection, experience, evidence, sensitivity, knowledge, and skill, and that ethics is sensitive to the values and contexts in which various questions arise.

ACTIVITY FOUR

CLOSING TEASER— TOMORROW'S HEADLINE

Time needed for completion:

5 minutes

MATERIALS

For the class:

- Overhead projector
- Transparency P

For each student:

- Lesson 7 homework extension

Procedure

After completing the other activities, assign the homework extension, then display Transparency P: *Researcher Seeks Approval for Experiment* without providing additional details.

Homework Extension: You Make the Call

Hand out Homework Extension: *You Make the Call* (p. 138) for students to complete before the next lesson. Remind students to think about the key ethics terms as they complete the assignment.

Optional Extension

To integrate the unit as part of an interdisciplinary approach with a mathematics course, have students develop an ethics survey independently or in cooperative groups. The students develop and conduct the survey with other students, friends, and family members about their attitudes and knowledge of ethics based on a numerical scale. Following the survey, the students could statistically analyze and graph their results and present their findings to the class.

Resources for Further Exploration

Bebeau, M. J., J. R. Rest, and D. Narvaez. 1999. Beyond the promise: A perspective on research in moral education. *Educational Researchers* 28 (4): 18–26.

Benninga, J. S., M. W. Berkowitz, P. Kuehn, and K. Smith. 2003. The relationships of character education and academic achievement in elementary schools. *Journal of Research in Character Education* 1 (1): 17–30.

Character Education Partnership. 2002. *Practices of teacher educators committed to characters. Examples from teacher education programs emphasizing character development.* Washington, DC: Character Education Partnership.

Goleman, D. 1995. *Emotional intelligence: Why it can matter more than IQ.* New York: Bantam.

The Online Ethics Center for Engineering and Science at Case Western Reserve University. *Ethics in the science classroom: Introduction. www. onlineethics.org/cms/9659.aspx*

Roberts, L. W., and A. R. Dyer. 2004. *Ethics in mental health care.* Washington, DC: American Psychiatric Publishing.

Ryan, K., and K. E. Bohin. 1999. *Building character in schools: Practical ways to bring moral instruction to life.* San Francisco, CA: Jossey-Bass.

Shapiro, D. A. 1999. Teaching ethics from the inside-out: Some strategies for developing moral reasoning skills in middle-school students. Paper presented to the Seattle Pacific University Conference on the Social and Moral Fabric of School Life, Edmonds, WA.

WHAT IS ETHICS IN SCIENCE?

TRANSPARENCY O

Key Ethics Terms

Key Term	Definition	Example
Autonomy	A person's ability to make his or her own decisions.	
Beneficence	Acting in a way that benefits people or animals; doing good for others.	
Compassion	Genuine care for the suffering of others including providing kindness and comfort.	
Justice	Treating others fairly, and having the right to equal opportunities.	
Nonmaleficence	Intentionally trying to not harm others, and preventing harm from occurring to people or animals.	
Veracity	Being truthful	

WHAT IS ETHICS IN SCIENCE?
TRANSPARENCY P

Name_____ Date_____

WHAT IS ETHICS IN SCIENCE?
WORKSHEET 13

What does the word *ethics* mean to you?

_____

_____

_____

Key Ethics Terms

Key Term	Definition	Example
Autonomy	A person's ability to make his or her own decisions.	
Beneficence	Acting in a way that benefits people or animals; doing good for others.	
Compassion	Genuine care for the suffering of others including providing kindness and comfort.	
Justice	Treating others fairly, and having the right to equal opportunities.	
Nonmaleficence	Intentionally trying to not harm others, and preventing harm from occurring to people or animals.	
Veracity	Being truthful	

Name_____ Date_____

WHAT WOULD YOU DO?
WORKSHEET 14

Introduction

As a class, read "To Tell or Not to Tell," below. In your small groups, complete the questions based on the Key Ethics Terms from Worksheet 13. Discuss the story with your group and record your answers in the spaces provided. Remember, there will not be one "right" answer because people have different opinions about how they would react to the situation, and why. Even if everyone cannot agree on what is the best response, ethically, it is sometimes clear what a "wrong" thing might be. If members in your group do not agree, try to explain why. Look for things you agree on, as well as those you don't agree on within your group and the class.

To Tell or Not to Tell ...

You are friends with Chris. You noticed lately that when you're hanging out or playing baseball, he seems to always be tired. In fact, for the past few weeks, every time you call him to get together, he says that he'd rather stay home. A few weeks ago you even noticed that he was sleeping in class. When you asked him about what's going on lately, he said that he was probably staying up too late playing video games. Now that something serious has happened to Chris, you're wondering if you should have said something to Chris's mom, his brother, or your teacher about the change in his behavior.

Questions for Exploration

1. **What ethical issues are raised in this situation?**

_____
_____
_____
_____
_____
_____
_____
_____
_____
_____
_____

2. **How did your group respond? Should you have talked to someone about Chris?**

_____
_____
_____
_____
_____
_____
_____
_____
_____
_____
_____

Name_____ Date_____

WHAT WOULD YOU DO?
WORKSHEET 14 (CONTINUED)

3. **Explain how the following ethical concepts relate to this situation.**

Beneficence_____
_____

Compassion_____
_____

Non-maleficence_____
_____

Veracity _____
_____

4. **As a group, create your definition for "ethics" and write it below:**

_____
_____
_____
_____
_____
_____
_____
_____
_____
_____
_____
_____
_____
_____
_____

Name_____ Date_____

YOU MAKE THE CALL
HOMEWORK EXTENSION

1. Read the following situation and complete the questions using the same process that was modeled in class.

Payton has been doing an experiment in his seventh-grade science class to calculate the density of copper. In the experiment, he determines the mass of a penny and then drops it into a graduated cylinder containing water to determine the volume of the penny. When Payton plots his data (mass vs. volume) on a graph, he notices that all of the points fall on the line, except one. Payton assumes he must have made an error when measuring the mass or the volume of the penny. He has been worried about his lab grade and is concerned that if he turns in the lab report with the data he collected, he will receive a poor score. Payton is considering erasing the data that does not fall on the line from his data table and graph, so that the remaining results appear correct.

a. What is the question Payton is struggling with?

_____
_____
_____
_____
_____

b. What are his options and how do you see each of them working out?

_____
_____
_____
_____
_____
_____

YOU MAKE THE CALL
HOMEWORK EXTENSION (CONTINUED)

c. Explain how the following ethical concepts relate to this situation:

Veracity _____

_____

Nonmaleficence _____

_____

d. What might you do in this situation and why?

_____
_____
_____
_____
_____
_____
_____
_____
_____
_____
_____
_____
_____
_____
_____
_____
_____
_____

LESSON 8

APPLYING ETHICS TO RESEARCH

Teacher Lesson Plan: Overview and Concepts

In this lesson, students explore ethical considerations in research, focusing specifically on the example of the inclusion of animals in medical research. They work in small groups to analyze the possible harms and benefits to both humans and animals of two hypothetical medical research scenarios, both of which involve animal research. Students review each scenario with the goal of deciding which they would approve for study — one, both, or neither. Students examine their chosen scenario and list several factors/guidelines that will be important for a researcher to follow to ensure the animals are treated ethically.

Objectives

After completing this lesson, students will be able to

- work cooperatively in small groups to analyze information and draw conclusions;

- develop an understanding of what researchers need to consider when including animals in research;

- develop a deeper understanding of the contributions of animals in medical research;

- analyze the possible harms/benefits of scientific research to society and animals; and

- develop thoughtful questions that reflect the scientific inquiry process.

8 APPLYING ETHICS TO RESEARCH

National Standards Met in Lesson 8

National Science Education Standards

Standard A: Science as Inquiry

- Abilities necessary to do scientific inquiry
- Understanding about scientific inquiry

Standard C: Life Science

- Structure and function in living systems
- Regulation and behavior

Standard E: Science and Technology

- Abilities of technological design

Standard F: Science in Personal and Social Perspectives

- Personal health
- Risks and benefits
- Science and technology in society

Standard G: History and Nature of Science

- Nature of science

National Council of Teachers of English

- Students employ a wide range of strategies as they write and use different writing process elements appropriately to communicate with different audiences for a variety of purposes. (standard 5)
- Students use spoken, written, and visual language to accomplish their own purposes (e.g., for learning, enjoyment, persuasion, and the exchange of information. (standard 12)

National Health Education Standards

- Students will analyze the influence of technology on personal and family health. (standard 4)
- Students will demonstrate the ability to apply a decision-making process to health issues and problems individually and collaboratively. (standard 6)
- Students will be able to express information and opinions about health issues. (standard 7)

National Social Studies Standards

- Social studies programs should include experiences that provide for the study of relationships among science, technology, and society. (standard 8)

Background for Teachers

Animal models in research have historically played a pivotal role in the advancement of medical technology, disease prevention, and treatment methods. Our society has regulations that specify that research of importance to human disease should be conducted first with animals and then, if it is deemed to be safe, with human volunteers. In studies related to addiction and drug abuse, the inclusion of animals has proven to be very important in the investigation of the physiological and behavioral mechanisms of drug abuse, and the exploration of certain possible treatments. Nonetheless, inclusion of animals in research continues to be a topic of public and scientific debate.

In general, the historical contributions of animal research for the advancement of human welfare are well understood. However, over time, incidents of animal mistreatment have occurred in the name of scientific advancement. Because of these events, the concept of animal ethics took root in the late 1780s, grew with the passage of the first animal protection legislation known as the 1876 Cruelty to Animals Act in England (which gave rise to the antivivisection movement), intensified in the 1970s with the development of the modern animal rights movement, and continues today. At the center of the debate are questions of whether the inclusion of animals in research is ethical and acceptable, how research use of animals resembles or differs from other uses of animals in our society, the degree to which animals have moral standing, and the extent to which people have the right to "use" animals for benefit to humanity.

Ideally, a far-reaching and important ethical issue of this nature should be addressed through a systematic and careful analysis—one that includes logical arguments, a reflection on values, and the perspectives and experiences of individuals throughout society. Strong emotional responses typically do not lead to resolution of complex moral dilemmas. In students, however, passionate responses are important because they reflect their emerging strengths, courage, and values-defining thoughts. Teaching about ethics in science is challenging! It is important for students to have an opportunity to discuss their opinions, which helps them shape a deeper understanding of the issue and develop an informed position.

Among the positive aspects of this ongoing debate, discussions have led to improvements in the care of laboratory animals. For example, as we'll discuss in Lesson 9, the Animal Welfare Act (AWA) and Institutional Animal Care and Use Committees (IACUC) have been established to oversee the treatment of animals. In addition, researchers around the world have become increasingly aware of the importance of caring for laboratory animals in an ethically acceptable and compassionate manner.

ACTIVITY ONE
ANIMALS IN MEDICAL RESEARCH

Time needed for completion:

20 minutes

MATERIALS

For the class:

- Overhead projector
- Transparency P

For each student

- Scenario #1 handout
- Scenario #2 handout
- Worksheet 15

Procedure

In this lesson, students will read and evaluate two hypothetical research scenarios that involve animals. They will discuss and compare the possible harms and benefits of these scenarios, then use their analyses to approve one, both, or neither scenario for study. Students must be able to explain their decisions.

1. Display Transparency P: *Researcher Seeks Approval for Experiment* from Lesson 7. Ask the students what they think this headline means. Engage students in a discussion about what a researcher might seek approval for, who must a researcher get approval from, and why approval is necessary.

2. Remind students of Transparency M: *Can a Mouse Help Chris?* from Lesson 5. Ask students if they think that scientists could include animals in research to help them better understand drug abuse and addiction and to develop treatment plans for patients such as Chris.

3. Explain to the students that, for the next exercise, they are members of a Review Board Committee at a biomedical research facility. (Clarify that *biomedical research* involves the application of the principles of the natural sciences, especially biology and physiology, to clinical medicine.) As Review Board Committee members, they are responsible for approving all research that happens at their facility. As part of this approval process, they must review all research *protocols* (plans) to ensure that they are scientifically sound and ethically acceptable. Last week, two research protocols involving animals were submitted for review. It is the students' job today to review each proposal and decide which protocol(s) they would approve for study—one, both, or neither. It is possible that the students may choose to deny approval for both. Explain that as a class they will evaluate the first scenario and then they will break

into their small groups to evaluate the second. Encourage students to critically evaluate each in terms of study design and ethical soundness based on the principles they have learned so far.

4. Distribute Scenario #1—*Evaluating Drug XYZ* (p. 150) to each student and read it together as a class. On the board make two T-charts, one for humans and one for animals. On each chart, label the left side "harms" and the right side "benefits."

5. As a class, discuss the potential harms and benefits for Scenario #1 (i.e., harms: animal life, cost of study and benefits: save human lives, improve animal care, discover new medications/cures). Have students complete the T-charts in Part A *only* on the first page of Worksheet 15: *What Are the Potential Harms? What Are the Potential Benefits?* (p. 152). Explain to students that no one wants to have animals euthanized, but in some circumstances it is necessary because there is no other reasonable option.

6. Once the class has finished evaluating Scenario #1, have them break into their small groups and hand out Scenario #2—*Nicotine Addiction: A Thing of the Past* (p. 151). Have the students repeat the same process to review Scenario #2 and complete the T-charts in Part A *only* on the second page of Worksheet 15: *What Are the Potential Harms? What Are the Potential Benefits?*

Two T-Charts for Scenario #1

Humans		Animals	
Harms	Benefits	Harms	Benefits

ACTIVITY TWO

DOES ETHICS APPLY TO ANIMALS?

Time needed for completion:

20 minutes

MATERIALS

For each student:

- Scenario #1 handout
- Scenario #2 handout
- Worksheet 15

Procedure

In this activity, students will revisit Scenarios #1 and #2 using the basic principles of ethics—autonomy, beneficence, compassion, justice, non-maleficence, and veracity—from Lesson 7.

1. Have the students work together in their small groups to once again review Scenarios #1 and #2 using the key ethics principles explored in Lesson 7. As the students work through each scenario from an ethics point of view, have them complete Part B of Worksheet 15: *What Are the Potential Harms? What Are the Potential Benefits?* for each scenario.

2. After students have completed Part B for both scenarios, have them come to a group consensus as to which scenario(s) they would approve for study—one, both, or neither. Students should justify their choices in Part D of Worksheet 15.

3. Engage the students in a class discussion about which research scenario(s), if any, their group decided to approve for study and why. Encourage the students to justify their responses. By reviewing the ethics principles and their application to these research studies, students should begin to raise questions about the need for more specific information in the research plan.

ACTIVITY THREE

CLOSING TEASER— TOMORROW'S HEADLINE

MATERIALS

For the class:

- Overhead projector
- Transparency Q

Procedure

After completing the other activities, assign the homework extension, then display Transparency Q: *Animal Care and Use Committee Alerted* without providing additional details or discussion.

Homework Extension: Animal Care and Medical Research

Hand out Homework Extension: *Animal Care and Medical Research* (p. 155) for students to complete before the next lesson.

Optional Extensions

- Using Scenario #1, have students develop a research plan structured around the scientific method that includes a scientific question, hypothesis, control groups, frequency of dosage, etc.

- Invite a researcher to visit your classroom to share her or his experience in biomedical research and the use of animals in research.

- Have students visit the Kids 4 Research website at *www.kids4research. org/teachers_parents/information.asp* to learn more about various careers in animal research. Have students prepare an informational brochure or newsletter on the computer to introduce a biomedical career of their choice.

Resources for Further Exploration

Websites

National Institute on Drug Abuse:
www.drugabuse.gov

Kids for Research:
www.kids4research.org

Foundation for Biomedical Research:
www.fbresearch.org

Massachusetts Society for Medical Research:
www.msmr.org

Animal Welfare Information Center, U.S. Department of Agriculture:
http://awic.nal.usda.gov

Periodicals

Bishop, L. J., and A. L. Nolen. 2001. Animals in research and education: Ethical issues. *Kennedy Institute of Ethics Journal* 11: 91–112.

DeGrazia, D. 1999. Animal ethics around the turn of the twenty-first century. *Journal of Agricultural and Environmental Ethics* 11: 111–129.

Books

Gluck, J. P., T. DePasquale, and F. B. Orlans. 2002. *Applied ethics in animal research.* West Lafayette, IN: Purdue University.

National Academy of Science. 1996. *Guide for the care and use of laboratory animals.* Washington, DC: National Academies Press.

Name_____ Date_____

Scenario #1

EVALUATING DRUG XYZ

Need for Research

Drug XYZ is one of the newest illicit drugs used by teenagers in the United States. Over the past year, law enforcement and health officials have become aware of its dangers due to the deaths of 11 teens who had heart attacks shortly after taking the drug. The deaths occurred in three different states. Drug XYZ is a pill, is inexpensive compared to other illicit substances, and is readily available. This combination is dangerous for teenagers who see this drug as affordable and easy to get, but are not aware that it can cause severe health problems.

Scenario #1 proposes to study Drug XYZ to learn more about its effects.

Specific Research Questions

1. Is Drug XYZ physically addictive or just recreational?

2. What effect does the drug have on the heart and other body systems?

3. What behavior changes happen while under the influence of the drug?

Animal Involvement

A total of 50 rats will be included:

- 25 rats will be given the drug dissolved in their water with an eyedropper every morning

- 25 rats will be given plain water with an eyedropper at the same time

Everything else will be the same for the two groups of rats: standard laboratory cages, feeding schedules, amount of time for sleeping, and so on.

Brief Study Protocol

All rats will be observed throughout the day to see how the drug affects behavior patterns. After two weeks, both groups of rats will be given plain water through an eyedropper, with no drug. Behavior will be observed for another two weeks to monitor for evidence of addiction/withdrawal. Then, all of the rats will be euthanized (killed quickly and as humanely as possible, but without anesthesia since this might affect the results) so that their brains and other organ systems can be examined for evidence of damage.

Scenario #2

NICOTINE ADDICTION: A THING OF THE PAST

Need for Research

Nicotine, the main drug in cigarettes and chewing tobacco, is one of the most commonly used drugs in the United States. According to the National Institute on Drug Abuse, in 2002, 30% of people in this country age 12 or older used tobacco products at least one time in a given month. According to the Centers for Disease Control, tobacco use is the nation's leading cause of preventable death. More than 440,000 people die each year from causes related to tobacco use.

Research that helps scientists better understand why nicotine is addictive may help to prevent or reduce addiction. In a previous study, researchers found that some people have a genetic variation that causes a decrease in the activity of an enzyme in the body that breaks down nicotine and its by-products. When the activity of this enzyme slows down, the person is less likely to become addicted to nicotine.[1] As a result of the study, a new drug was developed that may stop the function of this enzyme in normal people. Scenario #2 proposes to test the new drug.

Specific Research Questions

1. Does administration of the test drug in fact reduce this enzyme activity?

2. Do nicotine-addicted mice who have received this drug show less craving for nicotine?

3. What side effects does the drug have?

4. Are there any other behavior changes while taking the drug?

Animal Involvement

A total of 50 mice will be included. All of the mice will be given nicotine dissolved in their water every morning with an eyedropper for two weeks. Then, the mice will be trained to press a lever in their cages that will deliver more nicotine-laced water whenever they want it. Once the mice have become addicted to nicotine, they will be split into two groups.

*   25 mice will be given the enzyme-inhibiting drug dissolved in their water with an eyedropper every morning

*   25 mice will be given just plain water with an eyedropper at the same time

Everything else will be the same for the two groups of mice: standard laboratory cages, feeding schedules, amount of time for sleeping, and so on.

Brief Study Protocol

All mice will be observed throughout the day for two weeks to see how the drug affects behavior patterns (especially how often they seek out more nicotine-water). After two weeks, both groups of mice will be given plain water through an eyedropper, with no drug. Then, all of the mice will be euthanized (killed quickly and as humanely as possible, but without anesthesia since this might affect the results) so that their brains and other organ systems can be examined for evidence of damage.

---

1   NIDA. InfoFacts: Cigarettes and other tobacco products. *www.drugabuse.gov/publications/infofacts/cigarettes-other-tobacco-products*

Name_____ Date_____

WHAT ARE THE POTENTIAL HARMS?
WHAT ARE THE POTENTIAL BENEFITS?
WORKSHEET 15

Part A

Complete the following T-charts based on the class discussion.

Scenario #1—Evaluating Drug XYZ

Humans		Animals	
Harms	Benefits	Harms	Benefits

Part B

Select two of the following ethics concepts and explain how it relates to this situation.

Autonomy _____

_____

Beneficence _____

_____

Compassion _____

_____

Justice _____

_____

Nonmaleficence _____

_____

Veracity _____

_____

WHAT ARE THE POTENTIAL HARMS?
WHAT ARE THE POTENTIAL BENEFITS?
WORKSHEET 15 (CONTINUED)

Scenario #2—Nicotine Addiction: A Thing of the Past

Humans		Animals	
Harms	Benefits	Harms	Benefits

Part C

Select two of the following ethics concepts and explain how it relates to this situation.

Autonomy_____

_____

Beneficence_____

_____

Compassion_____

_____

Justice_____

_____

Nonmaleficence _____

_____

Veracity_____

_____

Name_____ Date_____

WHAT ARE THE POTENTIAL HARMS? WHAT ARE THE POTENTIAL BENEFITS?

WORKSHEET 15 (CONTINUED)

Part D

Directions: In your small group, agree on which scenario(s), if any, your group would approve for study, using scientific and systematic ethical reasoning, and justify your decision.

<table>
<tr><td>Which scenario(s) would you approve, if any?</td></tr>
<tr><td>___ Evaluating Drug XYZ</td></tr>
<tr><td>___ Nicotine Addiction: A Thing of the Past</td></tr>
<tr><td>___ Neither study</td></tr>
</table>

Explain your reasoning:

_____
_____
_____
_____
_____
_____
_____
_____
_____
_____

Is there any additional information that you believe should be included in the research plans to help you make a more informed decision?

_____
_____
_____
_____
_____
_____
_____
_____
_____
_____

ANIMAL CARE AND MEDICAL RESEARCH
HOMEWORK EXTENSION

Once a research project involving the inclusion of animals has been approved for study and before research can begin, a researcher must have his or her animal care plan approved by the Institution's Animal Care and Use Committee. This committee ensures that the research animals are properly cared for in an ethically acceptable way and are included in research only if necessary.

List at least five areas of animal care *you* would want to see addressed in a researcher's animal care plan (i.e., feeding schedule, staff training, and so on). Explain your answers in the space provided.

Area 1: _____

Why: _____

_____

Area 2: _____

Why: _____

_____

Area 3: _____

Why: _____

_____

Area 4: _____

Why: _____

_____

Area 5: _____

Why: _____

_____

LESSON 9
ENSURING THE ETHICAL CONDUCT OF RESEARCH

Teacher Lesson Plan: Overview and Concepts

Through the years, experiments that involve animal models have helped scientists, doctors, and veterinarians develop many new medical treatments—not only for humans but also for animals. Even though researchers continually look for ways to "reduce, replace, or refine" the degree of animal involvement, animals continue to play an important role in the advancement of modern medicine.

In this lesson, students explore the federal regulations and guidelines that have been established to help ensure the correct treatment of animals, such as the Animal Welfare Act (AWA). They discuss acceptable and unacceptable research activities and behaviors, then use this information to create a questionnaire that will help regulate the housing of animals in their middle school classrooms. In this lesson, students learn about shared decision making and compromise in addressing difficult issues.

Objectives

After completing this lesson, students will be able to

- work cooperatively in small groups to analyze information;

- develop an understanding of what researchers need to consider when including animals in research;

- understand and explain the role of federal guidelines such as the Animal Welfare Act;

- understand and explain the role of an Institutional Animal Care and Use Committee;

- think critically about the inclusion of animals in the middle school classroom; and

- develop a set of guidelines for animals in the middle school classroom.

9 ENSURING THE ETHICAL CONDUCT OF RESEARCH

National Standards Met in Lesson 9

National Science Education Standards

Standard A: Science as Inquiry
- Abilities necessary to do scientific inquiry
- Understandings about scientific inquiry

Standard F: Science in Personal and Social Perspectives
- Personal health
- Risks and benefits
- Science and technology in society

National Council of Teachers of English
- Students adjust their use of spoken, written, and visual language (e.g., conventions, style, vocabulary) to communicate effectively with a variety of audiences and for different purposes.

(standard 4)
- Students employ a wide range of strategies as they write and use different writing process elements appropriately to communicate with different audiences for a variety of purposes. (standard 5)
- Students participate as knowledgeable, reflective, creative, and critical members of a variety of literacy communities. (standard 11)
- Students use spoken, written, and visual language to accomplish their own purposes (e.g., for learning, enjoyment, persuasion, and the exchange of information). (standard 12)

National Social Studies Standards
- Social studies programs should include experiences that provide for the study of relationships among science, technology, and society. (standard 8)

ENSURING THE ETHICAL CONDUCT OF RESEARCH 9

Background for Teachers

When a researcher proposes to include animals in an experiment, his or her research plan must be approved by an Institutional Animal Care and Use Committee (IACUC). An IACUC typically consists of a group of doctors, scientists, and others who carefully review the proposal to ensure that the animals will be cared for in accordance with applicable regulations and guidelines.

Providing protection to animals is an important priority in the United States and around the world. There are numerous regulations and guidelines in place to help ensure their ethical treatment, such as:

- The United States Department of Agriculture (USDA) has established a number of federal programs governing the care of animals, among them the Animal Welfare Act. Enacted in 1966, the AWA provides guidelines for the handling, treatment, housing (including ventilation and lighting), and veterinary care of animals included in research. More information can be found at *http://awic.nal.usda.gov*

- Federal committees, such as the Committee on Animal Research and Ethics (CARE) and the Interagency Research Animal Committee (IRAC), have established guidelines for the protection of animals.

- The National Research Council has published a document titled *Guide for the Care and Use of Laboratory Animals* (2011), which contains recommendations for the evaluation of animal care. It addresses issues such as the lab environment, animal housing, veterinary care, and personnel qualifications.

As an added assurance, ethical scientists place a high priority on the three Rs: (1) *reducing* the number of animals included in research; (2) *replacing* animals with other models when possible; and (3) *refining* procedures to ensure the most humane treatment of animals. (More information on the three Rs is included in Lesson 10.)

The above guiding principles are a "work in progress." Efforts will continue as our society strives to achieve comprehensive protection for all subjects of research.

ACTIVITY ONE

ANIMAL CARE PLAN: WHAT NEEDS TO BE CONSIDERED?

Time needed for completion:

15 minutes

MATERIALS

For the class:

- Overhead projector
- Transparency Q

For each student:

- Worksheet 16

Procedure

1. Display Transparency Q: *Animal Care and Use Committee Alerted* from Lesson 8. Tell students that the Institutional Animal Care and Use Committee decided not to approve either research plan from Lesson 8 because several things were missing and were considered incomplete.

2. Have students break into their small groups and hand out Worksheet 16: *Animal Care Plan: What Needs to be Considered?* (p. 167). Using their Lesson 8 homework if necessary, have students work in groups to complete the worksheet. For each question, students need to think of examples of both acceptable and unacceptable ways that animals could be cared for in the research setting.

3. When groups have completed the worksheet, pick one group to share their "acceptable" responses and a different group to share their "unacceptable" responses for each question. Continue to do this until all groups have shared at least once.

4. As a class, discuss the difference between these two methods (acceptable/unacceptable) of caring for animals. Why might it be important to have rules that govern and regulate the care of animals included in research (accurate results, ethical reasons, etc.)? Try and lead the discussion to the idea that it is important to protect animals included in scientific research.

ACTIVITY TWO
TALKING POINTS

Time needed for completion:

10 minutes

MATERIALS

For each student:

• Talking Points handout

Procedure

1. Distribute one copy of the *Talking Points* handout (pp. 170–171) to each student.

2. Have all students read the material and discuss in small groups. After 10 minutes, discuss the material as a class and make sure the students understand the role of the Animal Welfare Act and the Institutional Animal Care and Use Committees.

ACTIVITY THREE

INSTITUTIONAL ANIMAL CARE AND USE COMMITTEES (IACUCS)

Time needed for completion:

15 minutes

MATERIALS

For each student:

• Worksheet 17

Procedure

1. Begin this activity with a discussion about what students have learned in this lesson's first two activities (e.g., there are many rules and regulations that govern the care of and inclusion of animals in research).

2. Distribute Worksheet 17: *Animal Inclusion in the Middle School Classrooms* (p. 169). Explain to students that it is their job to use what they have learned about the Animal Welfare Act and the Institutional Animal Use and Care Committees to create an application to be used in their own school if animals are housed in classrooms. Working in their small groups, the students will, in essence, become their school's own Institutional Animal Care and Use Committee and create an application that classrooms in their school should complete if they have and care for animals.

3. Have students break into their groups and complete the assignment. Walk around the room and monitor each group's progress. If they are having a hard time, use some of the questions below to stimulate discussion:

   • What benefits to students come from bringing an animal into the classroom?

   • What type of animal will be housed in the classroom? Why?

   • How will the animal(s) be acquired?

   • Where will the animal(s) be kept? Things to consider: housing, temperature, ventilation, lighting, and so on.

   • How often will the animal(s) be given food/water? Who will give the animal(s) food/water? Is there a set feeding/watering schedule? Who will ensure that this schedule is being followed?

- Will the animal(s) receive regular exercise? Who will exercise the animal(s)? Is there a set exercise schedule? Who will ensure this schedule is being followed?

- Who will take care of the animal(s) on weekends?

- Who will take care of the animal(s) on holidays and school breaks?

- Who will take care of the animal(s) during the summer?

- Do the animal(s) need veterinary care? If so, how often and who will provide this service?

Let students know that if they don't finish, they can complete the worksheet as homework.

ACTIVITY FOUR

CLOSING TEASER—TOMORROW'S HEADLINE

Time needed for completion:

5 minutes

MATERIALS

For the class:

- Overhead projector
- Transparency R

For each student:

- Worksheet 17

Procedure

After completing the activities, assign the homework extension and display Transparency R: *Chris Released From Hospital!* without providing additional details.

Homework Extension: Animal Inclusion in the Middle School Classroom

If it is not finished yet, have students complete Worksheet 17: *Animal Inclusion in the Middle School Classroom* as their homework assignment for Lesson 9.

Optional Extensions

- Have students read the editorial, "Ethics and Animal Research."[1] Have students use the article to make a chart listing the policies and committees that have been created to regulate the inclusion of animals in research.

- Divide the class into two groups. Let the students know that they will be debating the topic of animal inclusion in research. Both groups will be given time to research the topic and prepare their group's position. Group 1 will defend animal research, and group 2 will refute the inclusion of animals in research.

---

1 Fitzpatrick, A. 2003. *Journal of Laboratory Clinical Medicine* 141 (2): 89–90.

Resources for Further Exploration

Websites

NIDA for Teens:

http://teens.drugabuse.gov/facts/facts_brain1.asp

Understanding Animal Research:

www.understandinganimalresearch.org.uk

Foundation for Biomedical Research:

www.fbresearch.org/TwoColumnWireframe.
aspx?pageid=112

Committee on Animal Research and Ethics:

www.apa.org/science/leadership/care/index.aspx

Periodicals

Bishop, L. J., and A. L. Nolen. 2001. Animals in research and education: Ethical issues. *Kennedy Institute of Ethics Journal* 11: 91–112.

Dennis, J. U. 1997. Morally relevant differences between animals and human beings justify the use of animals in biomedical research. *Journal of the American Veterinary Medical Association* 210 (5): 612–618.

Fitzpatrick, A. 2003. Ethics and animal research. *Journal of Laboratory and Clinical Medicine* 141 (2): 89–90.

Gluck, J.P., and J. Bell. 2003. Ethical issues in the use of animals in biomedical and psychopharmacological research. *Psychopharmacology* 171: 6–12.

Page, G. G. 2004. The importance of animal research in nursing science. *Nursing Outlook* 52 (4): 102–107.

Rowen, A. N. 1997. The benefits and ethics of animal research. *Scientific American* 276 (2): 79.

Witek-Janusek, L. 2004. Commentary on the importance of animal research in nursing science. *Nursing Outlook* 52 (4): 108–110.

ENSURING THE ETHICAL CONDUCT OF RESEARCH

TRANSPARENCY R

ANIMAL CARE PLAN: WHAT NEEDS TO BE CONSIDERED?
WORKSHEET 16

Work in small groups to complete the worksheet. For each question, think of examples of both acceptable and unacceptable ways that animals could be cared for in a research setting.

1. **What type of animal will be included in the research scenario your class chose for Worksheet 15? Why?**

_____
_____
_____
_____
_____
_____
_____

2. **How will the animals be acquired?**

Acceptable:
_____
_____
_____
_____

Unacceptable:
_____
_____
_____
_____

3. **How and where will the animals be kept (consider housing, temperature, ventilation, lighting, etc.)?**

Acceptable:
_____
_____
_____
_____

Unacceptable:
_____
_____
_____
_____

4. **How often will the animals be given food/water?**

Acceptable:
_____
_____
_____
_____

Unacceptable:
_____
_____
_____
_____

Name_____ Date_____

ANIMAL CARE PLAN: WHAT NEEDS TO BE CONSIDERED?

WORKSHEET 16 (CONTINUED)

5. How often will the animals receive veterinary care?

Acceptable:

_____
_____
_____

Unacceptable:

_____
_____
_____

6. Will the animals receive exercise?

Acceptable:

_____
_____
_____

Unacceptable:

_____
_____
_____

7. How will the animals be treated during research phase?

Acceptable:

_____
_____
_____

Unacceptable:

_____
_____
_____

ANIMAL INCLUSION IN THE MIDDLE SCHOOL CLASSROOM
WORKSHEET 17

Guidelines for Animal Care in Middle School Classrooms

Part A

Your assignment is to use what you have learned from the *Talking Points* about the inclusion and care of animals in research to create guidelines for housing animals in a middle school classroom. List questions below that you believe must be thought about and answered before an animal can be brought into the classroom.

Sample Questions:

1. Why is it valuable to have animals in the classroom?

2. What type of animal(s) will be housed in the classroom? Why?

3. Where will the animals be kept?

4. Who will care for the animal(s) during the week? On weekends?

Part B

Below create *your* list of questions that should be answered and reviewed by the Middle School Animal Use and Care Committee in order to decide whether the needs of animals are being considered before they are brought into the classroom:

_____
_____
_____
_____
_____
_____
_____
_____
_____
_____
_____
_____
_____
_____
_____
_____
_____
_____
_____

Name_____ Date_____

TALKING POINTS

Animal Welfare Act (AWA)— United States Department of Agriculture (USDA)

The Animal Welfare Act was first passed in 1966 and is the federal law that governs the humane care, handling, treatment, and transportation of animals. Not only does this include animals in laboratories and dealers who sell animals to laboratories, it also includes dog and cat breeders, zoos, and circuses. The AWA covers all species of animals except birds, mice, and rats because of the costs necessary to implement the protections. It is not because of any characteristic of the animals themselves.

Institutional Animal Care and Use Committees

In 1985, the AWA was amended and required institutions doing biomedical research to form Institutional Animal Care and Use Committees. These committees are responsible for reviewing all research proposals that include animals before research is allowed to begin. They are also responsible for overseeing all institutional animal care and use programs.

Reviewing Research Protocols

If scientists, medical doctors, or veterinarians want to do research involving animals, they have to follow a specific set of rules that are governed by Institutional Animal Care and Use Committees.

- First, they have to write a research protocol (plan). The protocol must describe the research, explain the animal model to be used, explain why the animal model was chosen, and include a detailed description of procedures and how the animals will be included in the experiment. The protocol must also ensure that no one else has done the same research before so animals are not included unnecessarily.

- The research protocol is then given to the Institutional Animal Care and Use Committee for scientific and ethical review. After careful consideration and discussion, the committee decides whether to approve the protocol or not.

TALKING POINTS (CONTINUED)

Overseeing Animal Use and Care Programs

The Institutional Animal Care and Use Committee is the "eyes and ears" of the research institution and is responsible for making sure everything in the research facility is kept up-to-date. This ongoing governance function includes:

- making inspections of the animal housing space and laboratories

- reviewing and revising the program for the Humane Care and Use of Animals

- making sure that all procedures involving animals are performed in consultation with a veterinarian

- making sure that any pain involved in animal experimentation is minimized with analgesics

- making sure that **all** people working with animals are trained

Additional Information

The vast majority of animals included in research are bred specifically for testing purposes. In addition to providing ethical care to these animals to help ensure their quality of life, scientists are motivated to keep them well-fed, well-housed, and free of illnesses to help ensure reliable experiment results. Lab animals that are unhealthy, stressed, or frightened can lead to unreliable scientific outcomes.

Some animals, such as dogs and cats, can be acquired directly from the "death row" of animal pounds or purchased from USDA licensed and regulated dealers. These are known as "random-sourced" animals and are not preferred for use in research because of too many unknowns (e.g., variability in genetics, disease history, lack of training to work in a lab environment).

LESSON 10

THINKING ABOUT THE FUTURE OF RESEARCH

Teacher Lesson Plan: Overview and Concepts

Scientists continually look for new research methods to answer important questions. They also look for more efficient and more ethical ways of doing scientific work. Scientists now are continually looking for approaches that may provide alternatives to the inclusion of animals in research. By using non-animal models such as computer simulations, mathematical modeling, and in vitro methods (e.g., cell culture), they are able to *refine, reduce,* and in some cases, *replace* the use of animals in research. These are referred to in the scientific community as the *three Rs.* Some even consider a fourth R, that each scientist has the *responsibility* to consider the three Rs.

In this lesson, students look toward the future of research, specifically animal research. Students discover alternative methods to animal research and what role these methods might play in the future. Students come to realize that animals will continue to play an important role in the advancement of modern medicine even with the development of new research technologies.

Objectives

After completing this lesson, students will be able to

- work cooperatively in small groups to analyze information;
- explain the three Rs (reduce, refine, and replace) of animal inclusion in research;
- develop an understanding of why animal research will continue to be needed;
- describe various ways scientists reduce, refine, and replace animals included in research; and
- evaluate the benefits and limitations of non-animal research.

10 THINKING ABOUT THE FUTURE OF RESEARCH

National Standards Met in Lesson 10

National Science Education Standards

Standard A: Science as Inquiry

- Abilities necessary to do scientific inquiry
- Understanding about scientific inquiry

Standard C: Life Science

- Structure and function in living systems

Standard E: Science and Technology

- Abilities of technological design
- Understandings about science and technology

Standard F: Science in Personal and Social Perspectives

- Science and technology in society

National Council of Teachers of English

- Students employ a wide range of strategies as they write and use different writing process elements appropriately to communicate with different audiences for a variety of purposes. (standard 5)
- Students use spoken, written, and visual language to accomplish their own purposes (e.g., for learning, enjoyment, persuasion, and the exchange of information). (standard 12)

National Council for the Social Studies

- Social studies programs should include experiences that provide for the study of relationships among science, technology, and society. (standard 8)

ACTIVITY ONE
THE THREE RS

Time needed for completion:

25 minutes

MATERIALS

For the class:

- Overhead projector
- Transparency S
- One set of Three Rs research cards (for each small group)

For each student:

- Worksheet 18

Procedure

1. Display Transparency S: *Animal Research Reduced at Local Lab.* Present a hypothetical situation in which a local research facility reports a drop of 40% for the inclusion of dogs and cats in their research over the last 20 years.

2. Have students work in their small groups for about five minutes to brainstorm reasons why the inclusion of animals in this facility has declined. Have students record their answers.

3. While the students are brainstorming, draw the following table on a chalkboard/dry-erase board. This table describes the three Rs (refine, replace, and reduce) that scientists consider when including animals in research.

The Three Rs

Refine	Replace	Reduce
Alter tests so that an animal's pain or distress is decreased to the absolute minimum possible while keeping the data reliable and valid. (Pain management)	Use non-animal methods instead of animal methods in research or include animals with less sensory perception if non-animal methods are not possible.	Lower the number of animals included in a specific research project or use newer statistical techniques that may give similar data quality.
Examples:	Examples:	Examples:

4. After a few minutes, review the three Rs table with the class and ask each group to share its reasons why animal research has declined in the last 40 years. As each group provides a reason, have the class decide in which column the reason belongs (i.e., Does it help refine animal research so pain and distress are minimized; does it replace animal models used in research; or does it reduce the number of animals included in research?).

5. Once all groups have had a chance to share their ideas, explain to the class that the three Rs are an important part of scientific research today. Ethically sensitive scientists have a responsibility to consider and apply the three Rs. In fact, all scientists *must,* under the AWA, refine their research so that animal pain and distress is minimized. This is a federal mandate for all researchers who submit protocols for approval through an Institutional Animal Care and Use Committee.

6. Distribute Worksheet 18: *Thinking about the Future of Research* (p. 184). Students are going to explore current methods that are used to address the three Rs.

7. Next, pass out one set of research cards (pp. 180–183) to each group. The groups will have five minutes to discuss their research cards and tape them in the appropriate column on the board.

8. Once all groups have finished, review all research cards and their placement in the table. As a class, discuss the pros and cons of each method. As you review the information, remind students to record the responses on Worksheet 18.

9. Finally, ask the students if they think that at this time animals could be eliminated from medical research completely. Have them record and justify their answers in their worksheets.

ACTIVITY **TWO**

CHRIS—WHAT'S CHANGED, WHAT'S NEXT?

Time needed for completion:

20 minutes

MATERIALS

For the class:

- Overhead projector
- Transparency R

For each student:

- Worksheet 19

Procedure

1. Display Transparency R: *Chris Released from Hospital!* from Lesson 9.

2. Hand out Worksheet 19: *Chris—What's Changed, What's Next?* (p. 185).

3. As a class, recap what has happened to Chris throughout this unit. Create a timeline for Chris, lesson by lesson. *Hint:* Use the closing teasers (i.e., headlines) from each lesson (see #4 below).

4. Overview of closing teaser headlines to create a timeline:

    - Chris Collapses in Gym! Ambulance arrives in record time to transport Chris to hospital

    - Chris Treated at Emergency Room: Test results will be revealed tomorrow

    - Chris Tests Positive for Drugs! Chris asks medical team: "Am I in trouble?"

    - What's Wrong, Chris? Doctor prepares questions for medical interview

    - How Have Drugs Affected Chris's Brain?

    - Can Research Help Chris?

    - Can a Mouse Help Chris?

    - Doctor Questions Chris's Best Friend

    - Researcher Seeks Approval for Experiment

    - Animal Care and Use Committee Alerted

    - Chris Released From Hospital!

5. Next, in their small groups, have students answer the following questions:

    a. How has "your" Chris changed?

        i. Academically

        ii. Socially

        iii. With his or her family

        iv. Regarding his or her health

    b. What will Chris be doing next week? Next month? In one year? In five years?

    c. In what ways has research possibly helped Chris?

Optional Extensions

- Have students create a visual timeline for Chris based on the unit. Students could create the timeline on paper or using a computer program such as Microsoft Publisher or Word.

- Have each small group create a short play or skit based on the events Chris experienced throughout the unit. Emphasis should be placed on the information learned pertaining to substance abuse and the benefits of scientific research.

Resources for Further Exploration

Websites

Johns Hopkins University Center for Alternatives to Animal Testing:
http://caat.jhsph.edu
National Institute of Environmental Health Sciences:
www.niehs.nih.gov

Periodicals

Join Hands. 1999. Alternative research methods, refinement, reduction, replacement of animals needed in scientific research. 800-933-8288.
North Carolina Association for Biomedical Research. 2001. What's the point of bioscience research? 919-785-1304.

RESEARCH CARDS

Computer Simulations

Computer simulations can help scientists understand and simulate living systems without including animals or other organisms.

Magnetic Resonance Imaging (MRI)

MRI is a method of creating three-dimensional structural images of a patient's internal organs.

Mathematical Modeling

Mathematical modeling can help scientists understand and simluate living systems without including animals or other organisms.

Genetic Markers

A genetic marker is a piece of DNA (gene) with a known location and function whose inheritance patterns can be followed.

RESEARCH CARDS

Cell Culture

A cell culture is a type of "in-vitro" (i.e., "in glass") test. Animal or human cells are grown in the laboratory in petri dishes that contain the nutrients they need to grow. Scientists can then test how substances affect these cells.

Blood Work

A blood sample is drawn from a person's body allowing numerous laboratory tests to be conducted.

PET Scan

A PET scan is a medical imaging technique that produces a three-dimensional image or map of functional processes in the body.

Tissue Culture

A tissue culture is a type of "in-vitro" (i.e., "in glass") test. Animal or human tissues are grown in the laboratory in petri dishes that contain the nutrients they need to grow. Scientists can then test how substances affect these body tissues.

RESEARCH CARDS

Use of Anesthesia

Anesthesia is the use of medications to block the perception of pain. This allows humans or animals to undergo surgery and other procedures without the distress and pain they would otherwise experience. It is given before the pain begins and may cause loss of consciousness.

Use of Pain Medication

Pain medication is used to reduce the discomfort a human or animal feels. The effects of anesthesia wear off after surgery or other procedures.

Chemical Simulation

Chemical simulations can help scientists understand and simulate chemical reactions, including drug and medication effects, in living systems.

Mechanical Simulation

Mechanical simulation can help scientists understand and simulate mechanical processes (e.g., beating of the heart) in living systems.

RESEARCH CARDS

Non-Mammalian Models

Non-mammalian models include bacteria, insects, plants, and other simple life forms. These can be used in place of mammals in research to provide insights into living processes.

Replacement Technique

The replacement technique involves the use of plants and microorganisms in place of conscious living vertebrates in research.

Genetic Alterations

Genetic alteration involves manipulating genes in animals so that they quickly and more predictably react to various toxins. Gene-altered animals, such as mice, can reduce the time and number of animals required to see if a chemical is safe.

Name_____ Date_____

THINKING ABOUT THE FUTURE OF RESEARCH
WORKSHEET 18

In your small groups, use the information that you have learned to complete the following table:

The Three Rs

Refine		
Alter tests so that an animal's pain or distress is decreased to the absolute minimum possible while keeping the data is reliable and valid. (Pain management)	**Pros**	**Cons**
Examples: **Use of Anesthesia** **Use of Pain Medication**	**Reduces animal suffering which is more humane and improves the quality of science.**	**The anesthesia or pain medication may have an adverse affect on the animal.**

Replace		
Use non-animal methods instead of animal methods in research or include animals with less sensory perception if non-animal methods are not possible.	**Pros**	**Cons**
Examples: **Computer Simulations** **PET Scan** **Magnetic Resonance Imaging (MRI)** **Cell Culture** **Mathematical Modeling** **Genetic Markers** **Replacement Techniques** **Tissue Cultures** **Chemical and Mechanical Simulations** **Mathematical Modeling**	**Eliminates harm to animals to answer a scientific question. May decrease the financial cost of the research.**	**Can limit the reliability of the scientific data because a simulation is used to represent a living system.**

Reduce		
Lower the number of animals included in a specific research project or use newer statistical techniques that may give similar data quality.	**Pros**	**Cons**
Examples: **Blood Work Genetic Markers** **Non-Mammalian Models** **Genetic Alterations** **Mathematical Modeling**	**Reduces/eliminates harm done to animals to answer the scientific question. May also decrease the financial cost.**	**May limit the reliability of the scientific data. There must be sufficient data for analysis.**

At this time, do you think that animals could be completely eliminated from medical research? Justify your response below.

_____

_____

_____

CHRIS—WHAT'S CHANGED, WHAT'S NEXT?
WORKSHEET 19

Review what has happened to Chris throughout this unit. Create a timeline for Chris, lesson by lesson. *Hint:* The newspaper headlines will help.

Lesson	What Happened to Chris?
1	
2	
3	
4	
5	
6	
7	
8	
9	
10	

Name_____ Date_____

CHRIS——WHAT'S CHANGED, WHAT'S NEXT?
WORKSHEET 19 (CONTINUED)

1. How has your Chris changed

a. Academically?

_____
_____
_____
_____
_____

b. Socially?

_____
_____
_____
_____
_____

c. With his or her family?

_____
_____
_____
_____
_____

d. Regarding his or her health?

_____
_____
_____
_____
_____

CHRIS—WHAT'S CHANGED, WHAT'S NEXT?
WORKSHEET 19 (CONTINUED)

2. What do you think Chris will be doing

   a.   next week?

_____

_____

_____

_____

_____

   b.   next month?

_____

_____

_____

_____

_____

   c.   in one year?

_____

_____

_____

_____

_____

   d.   in five years?

_____

_____

_____

_____

_____

Name_____ Date_____

CHRIS—WHAT'S CHANGED, WHAT'S NEXT?
WORKSHEET 19 (CONTINUED)

3. **Think about what's happened to Chris since the beginning of this course. Look at the following headlines and decide whether or not research helped Chris.**

   a. Chris Collapses in Gym! Chris Treated at Emergency Room.

   _____
   _____
   _____
   _____
   _____

   b. Chris Tests Positive for Drugs! Can a Mouse Help Chris?

   _____
   _____
   _____
   _____
   _____

   c. Chris Released From hospital.

   _____
   _____
   _____
   _____
   _____

4. **Write a paragraph that answers the following questions:**
   - Have your ideas about the importance of research changed?
   - How important is ethics in research?

   _____
   _____
   _____
   _____
   _____

SECTION THREE

ASSESSMENTS

Name_____ Date_____

THIS IS YOUR BRAIN:
TEACHING ABOUT NEUROSCIENCE AND ADDICTION RESEARCH
UNIT TEST

Part A: Multiple Choice

Directions: Write the letter of the best answer on the line provided.

_____ 1. **The word hypothesis came from a Greek word that means "to _____."**

    a. see
    b. test
    c. suppose
    d. conclude

_____ 2. **Animals have been included in biomedical research since the early _____.**

    a. 1600s
    b. 1700s
    c. 1800s
    d. 1900s

_____ 3. **The brain sends and receives messages through _____.**

    a. neurons
    b. protons
    c. neutrons
    d. electrons

_____ 4. **Which method of research would be best to answer the question, "Can using different rewards and consequences reduce cigarette use?"**

    a. Biological
    b. Behavioral
    c. Genetic
    d. Population-based

_____ 5. **What legislation protects animal inclusion in biomedical research?**

    a. Animal Welfare Act
    b. Ethical Code of 1966
    c. Institutional Animal Care and Use Act
    d. Interagency Research Animal Committee Act

_____ 6. **The gap between neurons is called the _____.**

    a. terminal
    b. axon
    c. myelin
    d. synapse

_____ 7. **People who start smoking before the age of 21 have the hardest time quitting.**

    a. True
    b. False

_____ 8. **A medical imaging technique that produces a 3-D image of functional processes in the brain is known as a/an _____.**

    a. x-ray
    b. MRI
    c. PET scan
    d. STAT

_____ 9. **Drugs of abuse stimulate the part of the brain that is responsible for _____.**

    a. feelings of appetite and drive
    b. coordination
    c. sight
    d. emotions

_____ 10. **Which of the following is NOT one of the three Rs of biomedical research?**

    a. Refine
    b. Reuse
    c. Reduce
    d. Replace

_____ 11. **Which method of research would be best to answer the question, "Is marijuana addictive?"**

    a. Genetic
    b. Behavioral
    c. Population-based
    d. Biological

_____ 12. **Uncontrollable and compulsive drug seeking and use, even in the face of negative health and social consequences, is known as drug _____.**

    a. seeking
    b. abuse
    c. addiction
    d. use

_____ 13. **Pain management for animals included in research is an example of _____.**

    a. rationing
    b. refinement
    c. reduction
    d. replacement

_____ 14. **A message travels across the synapse in the neuron as a/an _____ impulse.**

    a. chemical
    b. electrical
    c. magnetic
    d. mechanical

_____ 15. **What part of the neuron receives the message from another neuron?**

    a. Soma
    b. Axon
    c. Dendrite
    d. Synapse

_____ 16. **Doing good for other individuals, family members, and society as a whole is called _____.**

    a. justice
    b. autonomy
    c. nonmaleficence
    d. beneficence

_____ 17. **The neurotransmitter that produces feelings of pleasure when released by the brain's reward system is _____.**

    a. GABA
    b. serotonin
    c. dopamine
    d. ecstasy

_____ 18. **Rodents are the most common type of animal included in biomedical research.**

    a. True
    b. False

_____ 19. **Which of the following is not a principle of ethics?**

    a. Justice
    b. Maleficence
    c. Beneficence
    d. Nonmaleficence

_____ 20. **When someone uses a drug repeatedly, the person may need _____ to produce the same effect.**

    a. the same amount of the drug
    b. more of the drug
    c. less of the drug
    d. a different drug

Part B: Matching

Directions: Using the four terms listed, label the diagram below.

> **Axon terminal**
> **Synapse**
> **Receptor**
> **Neurotransmitter**

21. _____

The gap between the neurons.

22. _____

Located at many sites in the neuron,
mostly on dendrites. Receives the
chemical messages.

23. _____

Carries messages from one neuron
to another neuron.

24. _____

Place where messages are prepared
and launched to the next neuron.

National Institute on Drug Abuse

Part C: Short Answer

Directions: Write your answers below. Use the back of this sheet if you need more space.

25. Briefly describe the three Rs and how they apply to the inclusion of animals in research.

26. Define the term *ethics* and give an example of when you had to make an ethical decision. Which of the five principles of ethics apply to your example?

27. Describe how drugs of abuse can interfere with message transmission between neurons. Draw and label a diagram of one or more neurons as part of your answer.

28. Describe one example of how you have used the scientific method in the last month.

29. Why is drug addiction considered a brain disease?

Part D: Essay Question

Write your answers to both parts of the question below.

30. **Statistics show that the inclusion of some animals, such as cats and dogs, in biomedical research has decreased in the last few decades.**

    a. Explain how animals have contributed to advances in medical and veterinary care.

    b. Do you think there will ever be a time when animals will not play a role in biomedical research? Provide evidence or examples of why you feel that way.

Name_____ Date_____

THIS IS YOUR BRAIN:
TEACHING ABOUT NEUROSCIENCE AND ADDICTION RESEARCH
UNIT TEST

ANSWER KEY

Part A: Multiple Choice

Directions: Write the letter of the best answer on the line provided.

__C__ **1.** The word hypothesis came from a Greek word that means "to _____."

    a. see
    b. test
    c. suppose
    d. conclude

__A__ **2.** Animals have been included in biomedical research since the early _____.

    a. 1600s
    b. 1700s
    c. 1800s
    d. 1900s

__A__ **3.** The brain sends and receives messages through _____.

    a. neurons
    b. protons
    c. neutrons
    d. electrons

__B__ **4.** Which method of research would be best to answer the question, "Can using different rewards and consequences reduce cigarette use?"

    a. Biological
    b. Behavioral
    c. Genetic
    d. Population-based

__A__ **5.** What legislation protects animal inclusion in biomedical research?

    a. Animal Welfare Act
    b. Ethical Code of 1966
    c. Institutional Animal Care and Use Act
    d. Interagency Research Animal Committee Act

__D__ **6.** The gap between neurons is called the _____.

    a. terminal
    b. axon
    c. myelin
    d. synapse

A 7. **People who start smoking before the age of 21 have the hardest time quitting.**

a. True
b. False

C 8. **A medical imaging technique that produces a 3-D image of functional processes in the brain is known as a/an _____.**

a. x-ray
b. MRI
c. PET scan
d. STAT

A 9. Drugs of abuse stimulate the part of the brain that is responsible for _____.

a. feelings of appetite and drive
b. coordination
c. sight
d. emotions

B 10. **Which of the following is NOT one of the three Rs of biomedical research?**

a. Refine
b. Reuse
c. Reduce
d. Replace

D 11. **Which method of research would be best to answer the question, "Is marijuana addictive?"**

a. Genetic
b. Behavioral
c. Population-based
d. Biological

C 12. **Uncontrollable and compulsive drug seeking and use, even in the face of negative health and social consequences, is known as drug _____.**

a. seeking
b. abuse
c. addiction
d. use

B 13. **Pain management for animals included in research is an example of _____.**

a. rationing
b. refinement
c. reduction
d. replacement

A 14. **A message travels across the synapse in the neuron as a/an _____ impulse.**

a. chemical
b. electrical
c. magnetic
d. mechanical

C 15. **What part of the neuron receives the message from another neuron?**

a. Soma
b. Axon
c. Dendrite
d. Synapse

__D__ 16. **Doing good for other individuals, family members, and society as a whole is called _____.**

    a. justice
    b. autonomy
    c. nonmaleficence
    d. beneficence

__C__ 17. **The neurotransmitter that produces feelings of pleasure when released by the brain's reward system is _____.**

    a. GABA
    b. serotonin
    c. dopamine
    d. ecstasy

__A__ 18. **Rodents are the most common type of animal included in biomedical research.**

    a. True
    b. False

__B__ 19. **Which of the following is not a principle of ethics?**

    a. Justice
    b. Maleficence
    c. Beneficence
    b. Nonmaleficence

__B__ 20. **When someone uses a drug repeatedly, the person may need _____ to produce the same effect.**

    a. the same amount of the drug
    b. more of the drug
    c. less of the drug
    d. a different drug

Part B: Matching

Directions: Using the four terms listed, label the diagram below.

Axon terminal
Synapse
Receptor
Neurotransmitter

21. _____ **Synapse** _____

The gap between the neurons.

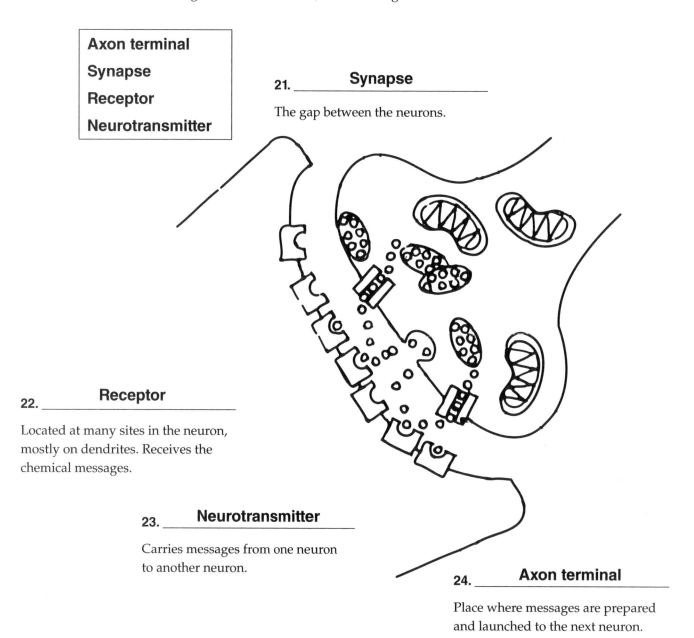

22. _____ **Receptor** _____

Located at many sites in the neuron, mostly on dendrites. Receives the chemical messages.

23. _____ **Neurotransmitter** _____

Carries messages from one neuron to another neuron.

24. _____ **Axon terminal** _____

Place where messages are prepared and launched to the next neuron.

National Institute on Drug Abuse

Part C: Short Answer

Directions: Write your answers below. Use the back of this sheet if you need more space.

25. **Briefly describe the three Rs and how they apply to the inclusion of animals in research.**

    Answer should include *Refine, Replace,* and *Reduce.*

    Answer should also describe the definitions:

    Refine - pain management;
    Replace - use of non-animals; and
    Reduce - lowering the number of animals included in research.

26. **Define the term *ethics* and give an example of when you had to make an ethical decision. Which of the five principles of ethics apply to your example?**

    Answer should be based on the class definition such as: Ways to understand and examine moral issues that are shaped by individual, community, and societal values and is arrived through a systematic process that generates acceptable, justifiable choices or options.

    Answer should also include a personal example as well as the correct ethics principle(s) that applies to the given example.

27. **Describe how drugs of abuse can interfere with message transmission between neurons. Draw and label a diagram of one or more neurons as part of your answer.**

    Answer should be based on material from Lesson 4, Activity 2.

28. **Describe one example of how you have used the scientific method in the last month.**

    Answers will vary.

29. **Why is drug addiction considered a brain disease?**

    Answer should be based on material from Lesson 4, Activity 2.

Part D: Essay Question

Directions: Write your answers to both parts of the question below.

30. **Statistics show that the inclusion of some animals, such as cats and dogs, in biomedical research has decreased in the last few decades.**

    a. Explain how animals have contributed to advances in medical and veterinary care.

    b. Do you think there will ever be a time when animals will not play a role in biomedical research? Provide evidence or examples of why you feel that way.

    Please see corresponding Unit Test Essay Question Assessment Rubric (next page).

Name_____ Date_____

THIS IS YOUR BRAIN:
TEACHING ABOUT NEUROSCIENCE AND ADDICTION RESEARCH
UNIT TEST ESSAY QUESTION ASSESSMENT RUBRIC

CATEGORY	Meets Standards	Approaching Standards	Below Standards	SCORE
Points Earned	3	2	1	
EVIDENCE AND EXAMPLES	Most of the evidence and examples are specific and relevant and explanations are given that show how each piece of evidence supports the author's position.	At least one of the pieces of evidence and examples is relevant and has an explanation that shows how that piece of evidence supports the author's position.	Evidence and examples are not relevant and/or are not adequately explained.	
POSITION STATEMENT	The position statement provides a clear statement of the author's position on the topic.	A position statement is present, but does not make the author's position clear.	There is no position statement.	
SENTENCE STRUCTURE	Most sentences are well-constructed and there is some varied sentence structure in the essay.	Most sentences are well-constructed, but there is little variation in structure.	Most sentences are not well-constructed or varied.	
FUNDAMENTALS	Very few or no minor errors in capitalization, punctuation, grammar, or spelling, and the essay is easy to read.	Some errors in capitalization, punctuation, grammar, or spelling which at times distracts the reader from the content.	Many errors in capitalization, punctuation, grammar, or spelling that distract the reader from the content.	

Essay Question Score: _____
12 possible points

Comments:

ALTERNATE ASSESSMENT OPTIONS

Board Game

Students can take aspects from the *This Is Your Brain* unit to develop an interactive board game. Remind students of the basics of board games: creative, colorful, and sturdy game board, unique playing pieces, dice or spinner, object of the game, how to win, clearly written instructions, fun, and so on. The game rules should be easy to follow, so it can be played without the creator being present. Stress the importance of quality content.

Upon completion of the board games, have students present their games to the class. Allow time for students to play each others' games and develop peer evaluation criteria.

Commercial

Have the students use facts that they have learned to make an informative antidrug message. Students could use Microsoft PowerPoint, audio, video, print, or any other type of media.

Newspaper

Using their project portfolios for content reference, have students work in their small groups to develop a school or class newspaper. Encourage students to use their creativity to write articles pertaining to the facts of drug use and abuse, addiction, how the brain functions and how drugs influence this process, the dangers of drug use and abuse, animal contributions to biomedical research, regulations guiding the inclusion of animals in research, letters to the editor, and so on. Students should write the articles as newspaper columnists and include visual aids (e.g., photographs, clip art, charts, graphs, tables).

Name_____ Date_____

POSTER PRESENTATION GUIDELINES AND ASSESSMENT RUBRIC

Based on the information compiled during the *This Is Your Brain* unit, design an informative and illustrative poster. The poster should be a minimum of 18″ × 24″ (maximum 24″ × 36″) and include a title and information on the following:

    a. What is drug addiction? How does addiction occur?

    b. What are some consequences of drug abuse?

    c. How are animals included in drug abuse research and what guidelines are in place to protect them?

The poster will be evaluated according to the following rubric:

CATEGORY	Excellent	Satisfactory	Needs Improvement	Unsatisfactory	SCORE
Points Earned	4	3	2	1	
FORMAT	Meets size requirements. Well-organized. Title gives specific information about main idea. Poster filled with appealing images and accurate information.	Meets size requirements. Good organization. Title gives some information about main idea. Poster mostly filled with appealing images and accurate information.	Does not meet size requirements. Somewhat difficult to read. Vague title. Fair organization. Poster somewhat filled with appealing images and accurate information.	Does not meet size requirements. Very difficult to read. Poster is missing a title. Poor organization. Little accurate information and few appealing images are presented.	

POSTER PRESENTATION GUIDELINES AND ASSESSMENT RUBRIC

CATEGORY	Excellent	Satisfactory	Needs Improvement	Unsatisfactory	SCORE
Points Earned	4	3	2	1	
CONTENT/ QUALITY OF INFORMATION	Topic and main ideas clearly stated. Includes specific and accurate supporting details from research. Shows depth of knowledge.	Topic and main ideas stated somewhat clearly. Includes most details from research. Shows good depth of knowledge. Information mostly accurate.	Topic and main ideas stated but not very clearly. Includes some details from research. Shows some depth of knowledge. Information somewhat accurate, somewhat complete.	Topic and main ideas vaguely stated. Includes few to no details from research. Shows little depth of knowledge. Information inaccurate or incomplete.	
VISUALS/ GRAPHICS	Includes exceptionally creative, colorful, and effective visuals. Clear labels. Visuals add strong support and interest to the content and purpose of the poster.	Includes some creative, colorful, and effective visuals. Good labels. Visuals add good support and interest to the content and purpose of the poster.	Includes few creative, colorful, and effective visuals. Fair labels. Visuals add some support and interest to the content and purpose of the poster.	Includes little to no visuals to support the content and purpose of the poster. No labels.	
MECHANICS	Very few or no errors in spelling, capitalization, punctuation, and usage.	Few errors in spelling, capitalization, punctuation, and usage.	Some errors in spelling, capitalization, punctuation, and usage.	Many errors in spelling, capitalization, punctuation, and usage.	

POSTER PRESENTATION GUIDELINES AND ASSESSMENT RUBRIC

CATEGORY	Excellent	Satisfactory	Needs Improvement	Unsatisfactory	SCORE
Points Earned	4	3	2	1	
NEATNESS	All images and information are presented in a very neat and visually attractive way. Exceptional presentation design. Very easy to read.	Images and information are neat and easy to interpret. Fairly effective presentation design. Easy to read.	Images and information are somewhat neat but difficult to interpret. Simple presentation design. Somewhat difficult to read.	Few to no images. Information is unclear and very difficult to interpret. Poor presentation design. Very difficult to read.	
TIMELINESS	Turned in on time.	Turned in one day late.	Turned in two days late.	Turned in three or more days late.	
				TOTAL SCORE	/24

Comments:

PROJECT PORTFOLIO COVER PAGE
HOMEWORK EXTENSION

Directions: Use the information from Worksheet 1 to create a portrait of Chris.

Answers will vary.

WHO IS CHRIS?
WORKSHEET 1

Part A—Group

Fill in the blanks to create a profile of Chris.

Age _____ Answers will vary. _____

Gender _____ Answers will vary. _____

Height _____ Answers will vary. _____

Hair color _____ Answers will vary. _____

Favorite book _____ Answers will vary. _____

Favorite song _____ Answers will vary. _____

Grade point average _____ Answers will vary. _____

Favorite food _____ Answers will vary. _____

Favorite sport _____ Answers will vary. _____

Favorite class _____ Answers will vary. _____

Favorite color _____ Answers will vary. _____

WHO IS CHRIS?
WORKSHEET 1 (CONTINUED)

Part A—Group (continued)

Complete these sentences.

Chris's personality can be described as Answers will vary.

Chris's dream job would be Answers will vary.

If Chris could change one thing about the world, it would be Answers will vary.

The best gift Chris ever received was Answers will vary.

Chris's family includes Answers will vary.

Chris's favorite science activity or experiment is Answers will vary.

Part B—On Your Own

Fill in the blanks to complete the activity.

A set of biographical details about a person is called a profile

The Greek word for "to suppose" is hypothesis

What question would you like to ask Chris that has not already been answered in class?

Answers will vary.

THE SCIENTIFIC METHOD
WORKSHEET 2

To be retained by students for reference

1. State the problem

What are you wondering about?

- Write the question(s) that the experiment will try to answer

2. Collect information

What do you already know?
What can you find out?

- Look for information about your question before you start your experiment
- How have others looked at this question?
- What will you measure?

3. Form a hypothesis

Based on what you know, what do you think will happen?

- Predict what the result of your experiment will show.
- Try to write your hypothesis as an "if/then" statement

4. Test your hypothesis

How will you conduct your experiment?

- Describe the steps of your experiment
- What materials will you need?

5. Observe and record your results

What happened in your experiment?

- Organize and record the data you collected

6. Draw a conclusion

Did you support your hypothesis?
What new questions do you have?

- State whether youR hypothesis was supported
- Share your conclusion with others

USING THE SCIENTIFIC METHOD
WORKSHEET 3

Student responses will vary based on scientific question.

1. State the problem
(What are you wondering about?)

2. Collect information
(What do you already know? What can you find out?)

3. Form a hypothesis
(Based on what you know, what do you think will happen?)

4. Test your hypothesis

(How will you conduct your experiment?)

5. Observe and record your results

(What happened in your experiment?)

6. Draw a conclusion

(Did you support your hypothesis? What new questions do you have?)

LOOKING AT THE FACTS
WORKSHEET 4

Part A

The categories in the table below show the areas of a person's life that can be negatively impacted by using drugs. Look at the table below and discuss the meaning of the data with your group.

A group of 10 middle school students were surveyed on how they thought drug use might most negatively impact areas of their lives. Students rated each topic with "1" indicating the area they felt would be least negatively impacted, "6" most negatively impacted.

| Category | Student ID number | | | | | | | | | | Total |
| | A | B | C | D | E | F | G | H | I | J | |
	Ratings										
Friendships	5	6	4	4	1	5	2	1	1	6	35
Grades	2	3	3	6	2	4	6	4	4	5	39
Family relationships	6	5	5	5	3	6	5	6	2	4	47
Sports performance	1	1	2	1	6	1	1	3	3	3	22
Health	4	2	6	3	4	2	3	2	5	2	33
Wake/sleep cycles (e.g., falling asleep in class, difficulty waking up in the morning)	3	4	1	2	5	3	4	5	6	1	34

1 = least impact; 6 = most impact

Part B

Decide what type of a graph you need.

- Line Graphs are used to track changes over time.
- Bar Graphs compare data between different groups or changes over time.

LOOKING AT THE FACTS
WORKSHEET 4 (CONTINUED)

Part C

Create your graph.

Many graphs have an x axis and a y axis. The x axis (a horizontal line) usually has numbers or labels for what is being measured, and the y axis (a vertical line) has numbers for the amount of the things being measured.

In the above table, what is being measured or compared? (*x* axis)

Areas of a student's life that are most negatively impacted by drug use.

Where can you find the measurement or comparison information? (*y* axis)

Ratings by students of which areas of life would be most negatively impacted by drug use.

Using graph paper, create a graph based on the information in the table on p. 50. Create your graph using a right angle with an x and y axis, labels for each axis, and points to represent your data. Also, don't forget to create a title for your graph.

Students should list areas of life on the x axis, and total student ratings on the y axis.

Part D

Based on the information discussed in this lesson, what is a new question that Chris's doctor might ask Chris?

Answers will vary.

USING THE FACTS
HOMEWORK EXTENSION

The graph below is based on a survey conducted by The National Household Survey on Drug Abuse in 2000. Students were asked about their use of cigarettes and illicit (illegal) drugs the month before the survey was taken. This graph shows the results of that survey.

Academic Performance and Use of Cigarettes or Illicit Drugs by Youth Age 12–17

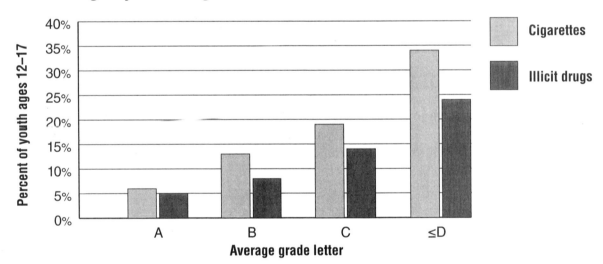

1. **How is each axis labeled in the graph above? Explain why.**

   x axis - Labeled with average grade letter because we are measuring the effect of drug use on a student's grade average.

   y axis - Percentages of drug-using students and their reported grade point averages

   How old were the students being studied?

   12–17 years

2. **What are the percentages by grade letter of reported illicit drug use in the students?**

   A = __5__ %   B = __9__ %

   C = __14__ %   ≤ D = __24__ %

USING THE FACTS
HOMEWORK EXTENSION (CONTINUED)

3. Students with grades of ≤ D used about ___28___ % more cigarettes and about ___19___ % more illicit drugs than students with a grade of A.

4. Write one conclusion based on the facts in this graph.

   Answers will vary. Answers should reflect the fact that there is a direct correlation between cigarette and/ or illicit drug use and poor academic performance.

The National Institute on Drug Abuse (NIDA) collected the following data in 1995 and again in 2003 on eigth-grade students who used illicit drugs:

Illicit Drug	Percent in 1995	Percent in 2003
Marijuana/Hashish	19.9	17.5
Inhalants	21.6	15.8
Hallucinogens and LSD	9.6	6.1
Cocaine and Crack Cocaine	6.9	6.1
Heroin	2.3	1.6
Steroids	2.0	2.5

5. Using graph paper (or the back of this page), draw a graph to show the difference in the percentage of drug use between 1995 and 2003. (Remember to include a title, axis labels, and measurements on your graph.)

6. Develop a one-sentence conclusion based on the facts presented in your graph.

   Answers will indicate that the use of most illicit drugs among 8th grade students decreased between 1995 and 2003 except for the use of steroids.

7. Write one question that you think the doctor will ask Chris based on the information provided.

   Answers will vary, but may include questions such as: "What drug have you been using?", "How long have you been using the drug?", "How often do you use the drug?", etc.

FUNCTIONS OF THE BRAIN
WORKSHEET 5

thought
speech
smell

touch
pain

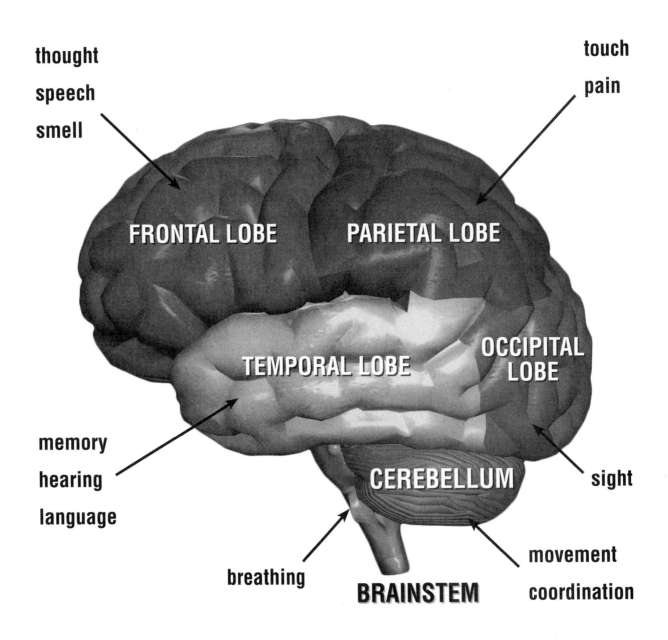

FRONTAL LOBE

PARIETAL LOBE

TEMPORAL LOBE

OCCIPITAL LOBE

memory
hearing
language

CEREBELLUM

sight

breathing

BRAINSTEM

movement
coordination

HOW THE NEURON SENDS A MESSAGE
WORKSHEET 6

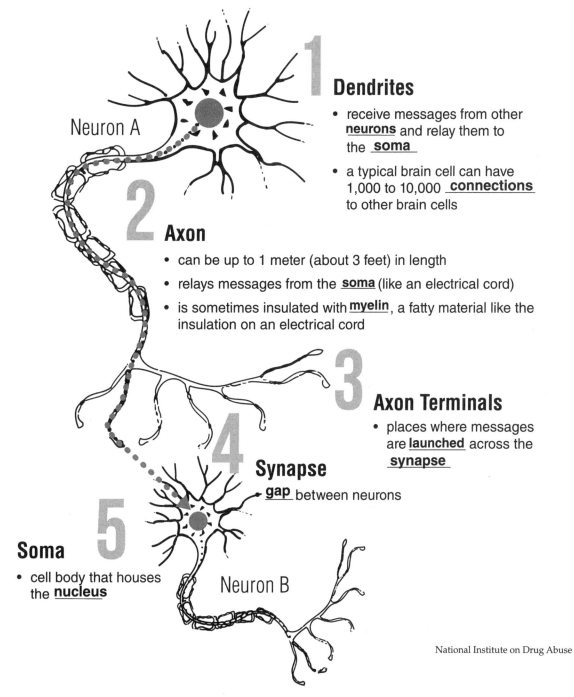

Neuron A

1

Dendrites
- receive messages from other **neurons** and relay them to the **soma**
- a typical brain cell can have 1,000 to 10,000 **connections** to other brain cells

2

Axon
- can be up to 1 meter (about 3 feet) in length
- relays messages from the **soma** (like an electrical cord)
- is sometimes insulated with **myelin**, a fatty material like the insulation on an electrical cord

3

Axon Terminals
- places where messages are **launched** across the **synapse**

4

Synapse
gap between neurons

5

Soma
- cell body that houses the **nucleus**

Neuron B

National Institute on Drug Abuse

SENDING THE MESSAGE ACROSS THE SYNAPSE
WORKSHEET 8

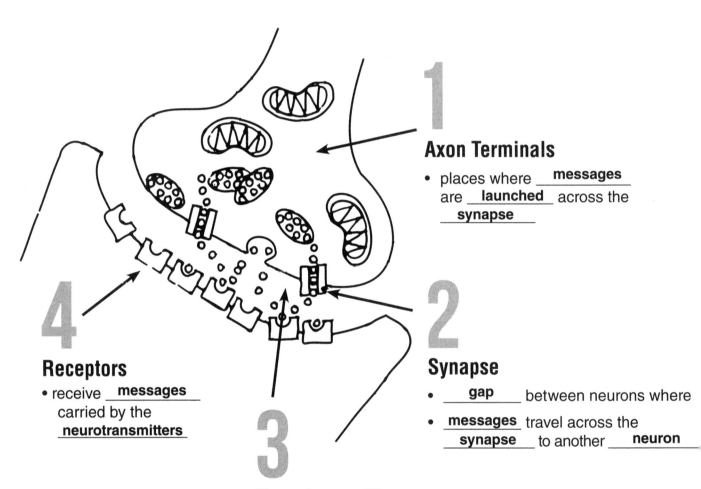

Axon Terminals
- places where __messages__ are __launched__ across the __synapse__

Synapse
- __gap__ between neurons where
- __messages__ travel across the __synapse__ to another __neuron__

Neurotransmitters
- carry __messages__ from one __neuron__ across the synapse to another __neuron__

Receptors
- receive __messages__ carried by the __neurotransmitters__

THE BRAIN GAME CROSSWORD PUZZLE
HOMEWORK EXTENSION

Use your Lesson 3 Worksheets to help solve the puzzle.

THE BRAIN GAME CROSSWORD PUZZLE
HOMEWORK EXTENSION (CONTINUED)

ACROSS

1. DENDRITES—Receive and relay message to the soma.

3. BRAINSTEM—You could not take a breath without this area of the brain.

4. SONNECTIONS—A typical brain can have thousands of these between brain cells.

7. CEREBELLUM—performing jumping jacks relies on this section of the brain.

9. SYNAPTIC—This gap is not a clothing store.

11. THOUGHT—The frontal lobe helps with this important job.

12. AXONTERMINAL—Messages are launched across the synapse from this location.

14. SOMA—The nucleus is housed in this cell body.

DOWN

2. TEMPORAL—This lobe makes hearing possible.

5. NEUROTRANSMITTERS—These carry messages across the synaptic gap.

6. RECEPTOR—Receives massages carried by neurotransmitters.

8. MYELIN—The axon is sometimes insulated with this material.

10. PAIN—One of the feelings sensed by the parietal lobe.

13. AXON—Messages from the soma are relayed through this part of the neuron.

PUTTING IT ALL TOGETHER

Based on the information from this crossword puzzle and the Lesson 3 classroom activities on the brain and neurons, summarize one new insight you have that might relate to Chris's condition.

> Responses will vary but should demonstrate a link between new information presented in Lesson 3 and Chris's situation.

BRAIN FUNCTION AND DRUG USE
WORKSHEET 9

Part A: Bridging the Gap

1. **Draw a detailed diagram of a two-neuron chain, including neurotransmitters, showing how a message is sent between neurons. Label the dendrites, soma, axon, axon terminals, synaptic gap, neurotransmitters, and receptors, and draw an arrow to indicate the direction of the electrical impulse. Be sure to show how the shape of the transmitter (object) fits the shape of the receptor.**

    Diagrams will vary. See Transparencies G and H for related drawings.

2. **Using the terms *message, neurotransmitters, lock, key, axon terminal, receptor*, and *synaptic gap*, explain how a message is transferred between neurons.**

    A message is carried through a neuron as an electrical message. At the axon terminal, the message is converted to a chemical message and is carried across the synaptic gap by a neurotransmitter. The neurotransmitter acts like a key that opens a lock at the receptor of the receiving neuron to receive the message. If the key matches the lock, the chemical message is delivered.

    (optional) If the key (neurotransmitter) does not match or it cannot find the lock (receptor), the message is lost.

BRAIN FUNCTION AND DRUG USE

WORKSHEET 9 (CONTINUED)

Part B: Breaking the Chain

3. Draw and label a colored diagram showing how a drug such as marijuana can interrupt message transfer between neurons. Be sure to include the axon terminal, synapse, neurotransmitters, and receptors.

    Drawings will vary. This is an example of how cocaine would affect message transfer.

    Note: This drawing is from: *http://www.teens. drugabuse.gov/mom/tg_brainimages_fig5.html*

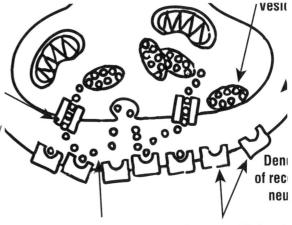

Neurotransmitters   Receptor Molecule

4. Write a paragraph to explain why drug addiction is considered a brain disease. Describe some ways in which drug use may have negatively affected Chris's brain and body.

    Answers will vary. Answer should include the definition of disease and a rationale for why drug addiction is a disease of the brain (i.e., drugs impair the normal functioning of the brain). See Lesson 4, page 4.7 for more information. Drug use may have affected Chris by: increased heart rate, respiratory problems, irritability, memory loss, etc.

5. Explain three ways a drug can interfere with natural message transmission between neurons.

    Drugs can interfere with normal message transmission by: 1) the drug may have a similar size and shape as the natural neurotransmitter, 2) the drug may block the reabsorption of the natural neurotransmitter, or 3) the drug may bind into the receptors and cause the neuron to release excess amounts of the neurotransmitter.

6. Do you believe scientific research can help Chris? Explain why or why not.

    Answers will vary.

THE BRAIN SCRAMBLER
HOMEWORK EXTENSION

Based on the material in your Project Portfolio and the Student Glossary (p. 212), follow the clues to unscramble each of these words and write the letters on the lines provided. Rearrange the shaded letters to solve the ultimate puzzle at the bottom of the page.

a. A short description of a person's characteristics

ILEROPF __ __ ▨ __ __ __ __   PROFILE

b. Neurotransmitter that produces feelings of pleasure when released by the brain's reward system

MOPIDENA __ __ __ __ __ ▨ __ __   DOPAMINE

c. Specialized branches that receive messages from other neurons and relay them to the soma

STENDDRIE __ __ __ ▨ __ __ ▨ __ __   DENDRITES

d. The addictive drug in tobacco

ENOCNTII __ __ ▨ __ __ __ __ __   NICOTINE

e. The repeated use of drugs to increase pleasure or reduce stress

GUDR SEAUB ▨ __ __ __ __ __ __ __ __
DRUG ABUSE

f. The part of the body that controls all thoughts, feelings, and other body functions.

NIRAB __ __ __ __ ▨   BRAIN

g. Part of the neuron that contains the nucleus

OAMS __ __ __ ▨   SOMA

h. A fatty material that surrounds and insulates the axons of some neurons

LYMNIE __ __ __ __ ▨ __   MYELIN

ULTIMATE PUZZLE:

Uncontrollable, compulsive drug seeking and use, even in the face of negative health and social consequences.

Answer: _____ ADDICTION _____

RESEARCH AND MARIJUANA USE
WORKSHEET 11

Activity 2

QUESTION: Do regular marijuana users develop breathing problems? (Study 18)

Fact: Regular marijuana users develop breathing problems including coughing and wheezing, and are more vulnerable to lung infections. Marijuana contains the same cancer-causing chemicals as tobacco. The amount of tar and carbon monoxide inhaled by marijuana smokers is three to five times greater than the levels of tobacco smokers.

WHO

Who were the subjects of study? *Hint:* There are two groups.

Marijuana users were considered the research subjects. Non-marijuana users were considered the control group (based on information in Lesson 5, page 5.5.)

WHAT

What did the researchers study?

Research Category: Biological

What are some research questions you would ask?

Answers will vary (based on information in Lesson 5.)

Student responses may include questions such as, "What are the long-term effects of marijuana use?" "How does marijuana affect the lung's ability to absorb oxygen?" or "How quickly do breathing problems occur when using marijuana?"

HOW

How could the researcher apply this information to benefit society?

Surgeon General's warnings, warnings on tobacco labels, improved health educational materials for children and adults, and advancements in medical treatments.

YOU MAKE THE CALL
HOMEWORK EXTENSION

Directions: Use Worksheets 10 and 11 as a guide to complete the following:

QUESTION: Are athletes more likely than non-athletes to abuse drugs?

WHO

Who would be the subjects of study? *Hint:* **There are two groups.**

Athletes were considered the research subjects. Non-athletes were considered the control group.

WHAT

What would YOU study?

Research Category: __Behavioral__

What data would you collect from each group?

Answers will vary and may include: type of physical activity involved in (organized/individual); amount of physical activity per week; ever used drugs (how often); type of drug used/tried; etc.

HOW

How might this information benefit other people?

Improved education for athletes and non-athletes, more effective treatments for substance abuse, improved understanding of the role of environmental factors, etc.

Based on the information you learned in this lesson, which category of research do you think would help Chris? Explain your answer.

Answers will vary, depending on the students' varying profiles of Chris.

ANIMALS AND MEDICAL DISCOVERIES
WORKSHEET 12 (CONTINUED)

1. **List two examples of how each animal or animal group listed below has contributed to advances in medical science:**

   Birds, Frogs, and Reptiles Edema treatments, malaria life cycle, social/behavioral patterns

   Cats Function of neurons, chemical transmission of nerve impulses, nerve function

   Cows, Sheep, and Pigs Blood transfusion, smallpox vaccine, parathyroid gland, MRI

   Dogs Blood transfusion, discovery of parathyroid gland, insulin and surgical techniques

   Monkeys, chimpanzees, etc. Rh factor in blood, chemotherapy, yellow fever vaccine

   Rabbits Cataract surgery, discovery of insulin, mechanism of diabetes

   Rodents Need for oxygen in respiration, discovery of penicillin, yellow fever vaccine

2. **Which type of animal do you think is most commonly included in medical research? Why?**

   Answers will vary.

3. **Why do you think animals are included in medical research?**

   Animals are included in research to learn more about biological systems and the diseases that affect both humans and animals. Through animal research, scientists gain a better understanding of various human and animal body systems and the diseases that afflict these systems so they are better able to advance medical treatments. Also because of ethical reasons, new treatments and drugs must first be tested on animals before humans because of possible dangerous effects. Animals are also included because they are more easily controlled in a laboratory environment.

ANIMALS AND MEDICAL DISCOVERIES
WORKSHEET 12 (CONTINUED)

4. Do you think animals also benefit from research? Can you think of any examples?

Animals are included in research to learn more about biological systems and the diseases that affect both humans and animals. Through animal research, scientists gain a better understanding of various human and animal body systems and the diseases that afflict these systems so they are better able to advance medical treatments. Also because of ethical reasons, new treatments and drugs must first be tested on animals before humans because of possible dangerous effects. Animals are also included because they are more easily controlled in a laboratory environment.

5. Can you think of ways in which animal research could be important in learning about drug abuse and addiction?

Animal research is beneficial in learning about drug abuse and addiction by helping scientists understand the biological and behavioral causes and consequences of drug use and abuse. For example, through animal research scientists are learning which areas of the brain are affected by various illicit drugs and how these drugs affect the human body as well as discovering more effective treatments and medications to help individuals with drug addiction.

WHAT DO YOU THINK?
HOMEWORK EXTENSION

Directions: Visit the website *www.kids4research.org*.
Use the information on these web pages to answer the following questions.

1. **Which type of animal is most commonly included in medical research? (List the percentages.) Why do you think this animal is included the most?**

    Most commonly used: Rats, mice, other rodents: 90%-95%

    Dogs, cats: Less than 1%

    Non-human primates: Less than 0.3%

    Why? Rodents are small; they cost less; scientists can breed many different strains of mice to imitate different diseases.

2. **Why have mice been included in research to study cancer?**

    Rodents are small; they are inexpensive to buy and care for; scientists can breed many different strains of mice to imitate different diseases and conditions. Cancer tumors can be placed in mice without rejection. Scientists can test new treatments on mice without risking human lives.

3. **How have rabbits been included in research to study drug abuse and addiction?**

    Rabbits have been included in research to study the effect of marijuana on the central nervous system.

4. **Which animals have been helpful to the sudy of AIDS**

    Mice, cats, and non-human primates?

WHAT DO YOU THINK?

HOMEWORK EXTENSION (CONTINUED)

5. **How have domesticated cats and dogs benefited from animal research?**

   Research including cats has led to better surgical procedures for animals and new vaccines. Research including dogs has led to the invention of devices for animals, including pacemakers and joint replacements, and has also resulted in better dental care for dogs, treatment for canine diabetes, and new vaccines.

6. **How have other types of animals helped medical science?**

   Ferret _____ improved our knowledge of influenza (the flu) _____

   Chinchilla _____ used to study middle ear infections in humans _____

   Armadillo _____ used to test a vaccine for leprosy _____

   Lobster _____ used to study Parkinson's disease, syphilis, and Huntington's disease _____

   Opossum _____ used to study the esophagus and bacterial endocarditis _____

7. **Visit the website _www.mismr.org/educational/drugabuse.html_. Use the information contained within these web pages to summarize one way in which animal research will help scientists and health care professionals learn more about drug abuse and addiction.**

   Future research may halp us learn more about hte effects of alcohol, tobacco, heroin, and cocaine.

   Answers will vary.

WHAT IS ETHICS IN SCIENCE?
WORKSHEET 13

What does the word *ethics* **mean to you?**

Answers will vary depending on the student's prior knowledge.

Key Ethics Terms

Key Term	Definition	Example
Autonomy	A person's ability to make his or her own decisions.	In research, this includes making sure people are fully informed before they consent to participate in a study and also to protect subjects who are not able to give informed consent.
Beneficence	Acting in a way that benefits people or animals; doing good for others.	In research, this includes designing studies and carrying them out in ways that should minimize harms and maximize benefits.
Compassion	Genuine care for the suffering of others including providing kindness and comfort.	Compassion involves being aware of those suffering and trying to relieve this suffering as much as possible with mercy and tenderness.
Justice	Treating others fairly, and having the right to equal opportunities.	In research, this includes making sure the benefits and burdens of the study are distributed fairly between the subjects and those who will benefit from the results.
Nonmaleficence	Intentionally trying to not harm others, and preventing harm from occurring to people or animals.	In research, this includes paying special attention to anything that causes physical or emotional suffering and making sure the anticipated benefits of the research are greater than the costs of doing the research.
Veracity	Being truthful	In research, volunteers must always be told the whole truth about an experiment, and they must be told in words that they understand.

*Sources: Reich, W.T. (Ed.). (1995). *Encyclopedia of bioethics.* New York: Macmillan Publishing Company.

Roberts, L.W. & Dyer, A.R. (2004). *Ethics in mental health care.* Washington, D.C.: American Psychiatric Publishing, Inc.

WHAT WOULD YOU DO?

WORKSHEET 14

Introduction

As a class, read "To Tell or Not to Tell," below. In your small groups, complete the questions based on the Key Ethics Terms from Worksheet 13. Discuss the story with your group and record your answers in the spaces provided. Remember, there will not be one "right" answer because people have different opinions about how they would react to the situation, and why. Even if everyone cannot agree on what is the best response, ethically, it is sometimes clear what a "wrong" thing might be. If members in your group do not agree, try to explain why. Look for things you agree on, as well as those you don't agree on within your group and the class.

To Tell or Not to Tell ...

You are friends with Chris. You noticed lately that when you're hanging out or playing baseball, he seems to always be tired. In fact, for the past few weeks, every time you call him to get together, he says that he'd rather stay home. A few weeks ago you even noticed that he was sleeping in class. When you asked him about what's going on lately, he said that he was probably staying up too late playing video games. Now that something serious has happened to Chris, you're wondering if you should have said something to Chris's mom, his brother, or your teacher about the change in his behavior.

Questions for Exploration

1. **What ethical issues are raised in this situation?**

   Responses may vary but should include student's struggle with not wanting to be responsible for getting a friend in trouble, but also being responsible enough to go to people who can help a friend when he's in trouble

2. **How did your group respond? Should you have talked to someone about Chris?**

   Responses will vary but each group should discuss differing opinions or comment on the group process in reaching a decision

3. **Explain how the following ethical concepts relate to this situation.** sample responses -

WHAT WOULD YOU DO?

WORKSHEET 14 (CONTINUED)

Beneficence Think about what was best for Chris instead of just thinking about how an action might affect you.

Compassion You would show compassion for Chris by talking to him about the changes in his behavior, and that you're concerned about why he's changed.

Nonmaleficence Sometimes it's difficult to understand why people behave the way they do, but if you think something harmful may be going on, ignoring it might cause more harm to the person.

Veracity Always be honest with your friends and others.

4. **As a group, create your definition for "ethics" and write it below:**

Responses will vary.

Be sure groups refer to key ethics terms in their definition.

YOU MAKE THE CALL
HOMEWORK EXTENSION

1. **Read the following situation and complete the questions using the same process that was modeled in class.**

   Payton has been doing an experiment in his seventh-grade science class to calculate the density of copper. In the experiment, he determines the mass of a penny and then drops it into a graduated cylinder containing water to determine the volume of the penny. When Payton plots his data (mass vs. volume) on a graph, he notices that all of the points fall on the line, except one. Payton assumes he must have made an error when measuring the mass or the volume of the penny. He has been worried about his lab grade and is concerned that if he turns in the lab report with the data he collected, he will receive a poor score. Payton is considering erasing the data that does not fall on the line from his data table and graph, so that the remaining results appear correct.

   a. **What is the question Payton is struggling with?**

   Payton is doing an experiment in his 7th grade science class. He is calculating the density of copper. For the experiment, Payton measured and graphed mass vs. volume. One data point did not fall on the line. Payton is worried about his lab grade and is thinking about changing his results. Payton wants to get good grades in science and he assumes he will get a poor grade on his lab report if one data point is incorrect. The ethical concern is that Payton is considering changing his lab results (i.e., falsifying data, cheating) so he does not receive a poor grade on his lab report.

   b. **What are his options and how do you see each of them working out?**

   Payton can change his results and present his report, and likely get a good grade. However, some people feel guilt when they act in a deceptive manner. Payton also has the option of presenting accurate results, describing the results in his report, and receiving a good score also. In this case, Payton would realize that sometimes things go wrong with experiments. Scientists look for causes of unusual results, learn from them, and discuss them in writing in order to share their experience with other scientists who may conduct a similar experiment.

YOU MAKE THE CALL
HOMEWORK EXTENSION (CONTINUED)

c. **Explain how the following ethical concepts relate to this situation:**

Veracity In order for Payton to display veracity, he must accept his true results, record them in his notebook, and accept his teacher's response.

Nonmaleficence By changing his results, Payton is intentionally being dishonest and hurting the other students in the class.

d. **What might you do in this situation and why?**

Student responses will vary, but should be based in, and reflect some key ethics principles.

WHAT ARE THE POTENTIAL HARMS?
WHAT ARE THE POTENTIAL BENEFITS?
WORKSHEET 15

Part A

Complete the following T-charts based on the class discussion.

Scenario #1—Evaluating Drug XYZ

Humans		Animals	
Harms	Benefits	Harms	Benefits
• Expense of the study	• Possibly save lives • Improve treatment options/medications • Greater understanding of how the drug affects humans	• Animal life • Potential pain/stress to the animals • Animals will be exposed to the drug	• Possibly improve animal treatments • Greater understanding of animal body systems

Part B

Select two of the following ethics concepts and explain how it relates to this situation.

Autonomy The animals included in the study should be cared for and housed in as natural an environment as possible to reduce stress and allow the animals to grow and remain healthy according to the type of animal included.

Beneficence The animals included in the study should be included in a way that will maximize the benefits (i.e., knowledge to be gained) while minimizing pain and suffering.

Compassion The animals should be cared for with kindness and a desire to eliminate as much suffering as possible.

Justice The study should make sure that the benefits and burdens of the study are distributed fairly.

Nonmaleficence The study should try to minimize physical or emotional suffering as much as possible and make sure the anticipated benefits of the research are greater than the costs of doing the research.

Veracity The research team must be honest in their report of how the animals will be treated, and then follow through with adherence to this standard.

WHAT ARE THE POTENTIAL HARMS?
WHAT ARE THE POTENTIAL BENEFITS?

WORKSHEET 15 (CONTINUED)

Scenario #2—Nicotine Addiction: A Thing of the Past

Humans		Animals	
Harms	Benefits	Harms	Benefits
• Expense of the study	• Possibly reduce the number of people addicted to or dying from nicotine • Improve treatment options/medications • Better understanding of how people become addicted to nicotine	• Animal life • Potential pain/stress to the animals • Animals will become addicted to nicotine and given the new drug	• Greater understanding of animal body systems

Part B

Select one of the following ethics concepts and explain how it relates to this situation.

Autonomy   Answers will vary but should be similar to Scenario 1.

Beneficence

Compassion

Justice

Nonmaleficence

Veracity

WHAT ARE THE POTENTIAL HARMS?
WHAT ARE THE POTENTIAL BENEFITS?

WORKSHEET 15 (CONTINUED)

Part D

Directions: In your small group, agree on which scenario(s), if any, your group would approve for study, using scientific and systematic ethical reasoning, and justify your decision.

Which scenario(s) would you approve, if any? Answers will vary.

____ Evaluating Drug XYZ

____ Nicotine Addiction: A Thing of the Past

____ Neither study

Explain your reasoning:

Answers will vary but they should be supported using the ethics principles previously discussed and not based on emotion.

Is there any additional information that you believe should be included in the research plans to help you make a more informed decision?

Additional information could include: How will the animals be cared for during the study? How often will they be fed? What will the living conditions be like for the animals? Will the study need to be repeated? Could the study be done with fewer animals or replaced with a more appropriate or non-animal alternative? How will pain and suffering be minimized? Why was this animal chosen for the study?

ANIMAL CARE AND MEDICAL RESEARCH
HOMEWORK EXTENSION

Once a research project involving the inclusion of animals has been approved for study and before research can begin, a researcher must have his or her animal care plan approved by the Institution's Animal Care and Use Committee. This committee ensures that the research animals are properly cared for in an ethically acceptable way and are included in research only if necessary.

List at least five areas of animal care *you* would want to see addressed in a researcher's animal care plan (i.e., feeding schedule, staff training, and so on). Explain your answers in the space provided.

Area 1: Feeding schedule

Why: To ensure the animals are being given water and appropriate food on a regular schedule.

Area 2: Housing specifications

Why: To ensure the animals are comfortable and free from as much environmental stress as possible for both ethical reasons and to improve the reliability of the results. Simulate a safe and natural environment.

Area 3: Training of the researchers/scientists/animal technicians

Why: Individuals working with animals in a research study should be appropriately trained about how to humanely handle and care for the animals to minimize pain and suffering.

Area 4: Health/veterinary care

Why: It is important for the animals to remain as healthy as possible throughout the study, again for ethical reasons and to improve the reliability of the results. Adequate veterinary care should be available.

Area 5: Where will the animals come from?

Why: The animals must come from a USDA licensed and regulated dealer for both ethical and scientific reasons.

ANIMAL CARE PLAN: WHAT NEEDS TO BE CONSIDERED?

WORKSHEET 16

Work in small groups to complete the worksheet. For each question, think of examples of both acceptable and unacceptable ways that animals could be cared for in a research setting.

1. **What type of animal will be included in the research scenario your class chose for Worksheet 15? Why?**

   Scenario 1: rats   Scenario 2: mice

2. **How will the animals be acquired?**

Acceptable:

   From a USDA licensed and regulated dealer.

Unacceptable: Steal the animals or buy them from an unlicensed facility or individual.

3. **How and where will the animals be kept (consider housing, temperature, ventilation, lighting, etc.)?**

Acceptable:

   The animals should be kept in a well ventilated and climate controlled lab meeting federal guidelines for temperature, lighting, number of animals per cage, appropriate space, etc. for rodents.

Unacceptable: It would be unacceptable to have too many animals per enclosure, little lighting, too hot or too cold of temperatures, little to no physical exercise, excessive noise, poor sanitation, etc.

4. **How often will the animals be given food/ water?**

Acceptable: The animals should be given adequate food/clean water for that animal species.

Unacceptable: Refraining from giving the animals food/clean water for extended periods of time if it is not relevant to the research plan.

ANIMAL CARE PLAN: WHAT NEEDS TO BE CONSIDERED?

WORKSHEET 16 (CONTINUED)

5. How often will the animals receive veterinary care?

Acceptable:

Acceptable veterinary care should be available for the animals and should be included in the research plan.

Unacceptable:

Not provide any or provide little care for animals that become ill.

6. Will the animals receive exercise?

Acceptable:

The animals should have regular exercise appropriate for the animal included in the study.

Unacceptable:

Have no physical or mental stimulation where the animals are housed.

7. How will the animals be treated during research phase?

Acceptable:

The animals should be treated humanely and with respect to the principles of ethics.

Unacceptable:

No attempt to minimize pain and suffering, inappropriate housing, lack of food/clean water, no veterinary care, etc.

ANIMAL INCLUSION IN THE MIDDLE SCHOOL CLASSROOM
WORKSHEET 17

Guidelines for Animal Care in Middle School Classrooms

Part A

Your assignment is to use what you have learned from the *Talking Points* about the inclusion and care of animals in research to create guidelines for housing animals in a middle school classroom. List questions below that you believe must be thought about and answered before an animal can be brought into the classroom.

Sample Questions:

1. Why is it valuable to have animals in the classroom?

2. What type of animal(s) will be housed in the classroom? Why?

3. Where will the animals be kept?

4. Who will care for the animal(s) during the week? On weekends?

Part B

Below create *your* list of questions that should be answered and reviewed by the Middle School Animal Use and Care Committee in order to decide whether the needs of animals are being considered before they are brought into the classroom:

Answers will vary but should reflect animal care discussed in this lesson.

THINKING ABOUT THE FUTURE OF RESEARCH

WORKSHEET 18 In your small groups, use the information that you have learned to complete the following table:

The Three Rs

Refine		
Alter tests so that an animal's pain or distress is decreased to the absolute minimum possible while keeping the data is reliable and valid. (Pain management)	**Pros**	**Cons**
Examples: Use of Anesthesia Use of Pain Medication	Reduces animal suffering which is more humane and improves the quality of science.	The anesthesia or pain medication may have an adverse affect on the animal.

Replace		
Use non-animal methods instead of animal methods in research or include animals with less sensory perception if non-animal methods are not possible.	**Pros**	**Cons**
Examples: Computer Simulations · PET Scan Magnetic Resonance Imaging (MRI) · Cell Culture Mathematical Modeling · Genetic Markers Replacement Techniques · Tissue Cultures Chemical and Mechanical Simulations Mathematical Modeling	Eliminates harm to animals to answer a scientific question. May decrease the financial cost of the research.	Can limit the reliability of the scientific data because a simulation Is used to represent a living system.

Reduce		
Lower the number of animals included in a specific research project or use newer statistical techniques that may give similar data quality.	**Pros**	**Cons**
Examples: Blood Work Genetic Markers Non-Mammalian Models Genetic Alterations Mathematical Modeling	Reduces/eliminates harm done to animals to answer the scientific question. May also decrease the financial cost.	May limit the reliability of the scientific data. There must be sufficient data for analysis.

At this time, do you think that animals could be completely eliminated from medical research? Justify your response below.

Answers will vary but should clearly indicate an understanding of the ethical guidelines involved when including animals in research as well as the application of the 3 Rs to research. Ethically, animals should only be included in research if there is no other way to answer the scientific question.

CHRIS—WHAT'S CHANGED, WHAT'S NEXT?
WORKSHEET 19

Review what has happened to Chris throughout this unit. Create a timeline for Chris, lesson by lesson. *Hint:* The newspaper headlines will help.

Lesson	What Happened to Chris?
1	Chris Collapses in Gym: Ambulance arrives in record time to transport Chris to hospital, and Chris Treated at Emergency Room: Test results will be revealed tomorrow
2	Chris Tests Positive for Drugs! Chris asks medical team: "Am I in trouble?"
3	What's Wrong, Chris? Doctor prepares questions for medical interview
4	Can Research Help Chris
5	Can a Mouse Help Chris?
6	Doctor Questions Chris's Best Friend
7	Researcher Seeks Approval for Experiment
8	Animal Care and Use Committee Alerted
9	Chris Released from Hospital
10	Animal Research Reduced at Local Lab

CHRIS—WHAT'S CHANGED, WHAT'S NEXT?

WORKSHEET 19 (CONTINUED)

1. **How has your Chris changed**

   a. Academically?

   > Answers will vary.

   b. Socially?

   > Answers will vary.

   c. With his or her family?

   > Answers will vary.

   d. Regarding his or her health?

   Answers could include that scientists have a better understanding of the science of drug dependence, a new treatment or medication was developed, etc.

2. **What do you think Chris will be doing**

   a. next week?

   > Answers will vary.

   b. next month?

   > Answers will vary.

   c. in one year?

   > Answers will vary.

   d. in five years?

   > Answers will vary.

CHRIS—WHAT'S CHANGED, WHAT'S NEXT?
WORKSHEET 19 (CONTINUED)

3. **Think about what's happened to Chris since the beginning of this course. Look at the following headlines and decide whether or not research helped Chris.**

   a. Chris Collapses in Gym! Chris Treated at Emergency Room.

   Responses can include points such as; emergency and hospital personnel are likely more prepared and knowledgeable about treating Chris and determining what's wrong with him due to past medical research studies. Diagnosis and treatment can start more quickly in emergency situations due to the knowledge gained from research

   b. Chris Tests Positive for Drugs! Can a Mouse Help Chris?

   Due to ongoing research, medical practitioners have a better understanding of the science of drug dependence, and of new treatment options. As discussed in earlier lessons, animals play an important role in research, and in particular, drug abuse research.

   c. Chris Released From hospital.

   Chris was properly diagnosed and treated due to discoveries made through medical and behavioral research. Effective treatment resulted in Chris's release from the hospital.

4. **Write a paragraph that answers the following questions:**

   • Have your ideas about the importance of research changed?

   • How important is ethics in research?

   Responses will vary but should display thoughtful reflection about the how the student's opinion or knowledge about research has changed, and about the importance of ethics in research and why ethics is important.

SECTION FOUR
SUPPORTING MATERIALS

MOUSE MAZE

Mouse Maze is a fun, interactive game designed to help students review the key concepts learned in the *This Is Your Brain* unit and help teachers assess students' understanding. Visit *www.nsta.org/publications/press/extras/brain.aspx* to download the game. Mouse Maze is available in both PC- and Mac-compatible formats.

STUDENT GLOSSARY

A

Addiction: An uncontrollable and compulsive drug seeking and use, even in the face of negative health and social consequences. A chronic, relapsing disease characterized by compulsive drug seeking and abuse and by long-lasting chemical changes in the brain. [1]

Amphetamines: A class of stimulant drugs that temporarily speed up body processes, causing the user to feel an increase in energy, alertness, and nervousness. [10]

Analgesics: A group of medications that reduce pain. [1]

Animal welfare: The philosophy that animals, especially those cared for by human beings, deserve proper care, including humane treatment, food, and shelter.

Animal Welfare Act (AWA): Federal law regulating the inclusion, sale, and handling of animals. [3]

Association for Assessment and Accreditation of Laboratory Animal Care (AAALAC): AAALAC International is a private, nonprofit organization that promotes the humane treatment of animals in science through voluntary accreditation and assessment programs. [2]

Autonomy: The right for a person to be independent, free from coercion, and able to make one's own decisions. [12,13]

Axon: The fiberlike extension of a neuron by which the cell carries information to target cells. [1]

Axon terminal: The structure at the end of an axon that produces and releases chemicals (neurotransmitters) to transmit the neuron's message across the synapse. [1]

B

Behavioral research: Behavior analysis is the scientific study of behavior. It measures observed behaviors, along with the role of environmental factors. Behavioral research helps us predict and control behavior. Studies have shown that behaviors are influenced by their consequences.

Beneficence: Commitment to acting in a way that brings about a benefit to people or animals, doing good for others. [12,13]

Bioethicist: An investigator who evaluates the ethical caliber of practices, procedures, and research within the life sciences and medical fields.

Bioethics: The study of the ethical and moral implications of practices, procedures, and research within the life sciences and medical fields.

Biological research: Biological research involves the study of the human body and all of its systems.

Biomedical research: Biomedical research involves: (a) the branch of medical science that studies the ability of humans to tolerate environmental stresses and variations, as in space travel, or (b) the application of the principles of the natural sciences, especially biology and physiology, to clinical medicine. [5, adapted]

Blood pressure: The pressure of the blood within the arteries.

Brain: The part of the central nervous system that is located within the cranium (skull). The brain functions as the primary receiver, organizer, and distributor of information for the entire body. It has two (right and left) halves called "hemispheres." [4, adapted]

Brainstem: The major route by which the forebrain sends information to, and receives information from, the spinal cord and peripheral nerves. [1]

C

Cell: The basic structural and functional unit in people and all living things. Each cell is a small container of chemicals and water wrapped in a membrane. There are 100 trillion cells in each of us, and each single cell in the human body contains the entire human genome, all the genetic information necessary to build a human being. [4, adapted]

Cell body (or soma): The central structure of a neuron, which contains the cell nucleus. The cell body contains the material that regulates the activity of the neuron. [1, adapted]

Central nervous system: The brain and spinal cord. [1]

Cerebellum: The section of the brain that helps to regulate movement and coordination.

Cerebral cortex: The outer layer of the brain that covers the two hemispheres of the brain. It is responsible for cognitive functions including reasoning, mood, and perception of stimuli. These are the "thinking cells." [1, adapted]

Cerebrum: The upper part of the brain consisting of the left and right hemispheres, each of which contains four separate areas: the frontal lobe, the parietal lobe, the occipital lobe, and the temporal lobe. [1, adapted]

Characteristic: A feature that helps to identify, tell apart, or recognize someone or something; a distinguishing mark or trait. [5]

Chemical message: A signal that is carried across a synapse from one neuron to another.

Chronic: Refers to a disease or condition that persists over a long period of time. [1]

Classify: To organize and arrange according to category.

Cocaine: A highly addictive stimulant drug derived from the coca plant that produces profound feelings of pleasure. [1]

Compare: To examine in order to note similarities between two items.

Compassion: The genuine regard for the suffering of another person or animal; includes a duty to respond to provide kindness and comfort for that individual. [12,13]

Computed tomography (CT) or computer axial tomography (CAT): An x-ray procedure that uses the help of a computer to produce a detailed picture of a cross section of the body. Also called a CT or CAT scan. [4]

Concept map: A type of diagram that shows various relationships between concepts. [6]

Conclude: To arrive at a judgment by the process of logical reasoning based on convincing evidence. [5]

Conclusion: A judgment reached by logical reasoning and convincing evidence.

Contrast: To examine in order to note the differences between two things. [5]

Control group: A group used as a standard of comparison in a controlled experiment. [5]

Craving: A powerful, often uncontrollable desire for a substance, often drugs. [1]

D

Data: A collection of facts from which conclusions may be drawn. [5]

Dendrite: The specialized branches that extend from a neuron's cell body to receive messages from other neurons. [1, adapted]

Disease: Illness or sickness often characterized by typical patient problems (symptoms) and physical findings (signs). [4]

Dopamine: A brain chemical, classified as a neurotransmitter, found in regions of the brain that regulate movement, emotion, motivation, and pleasure. [1]

Drug: A chemical compound or substance that can alter the structure of the brain and function of the body. Many drugs are illegal to use and possess. [1, adapted]

Drug abuse: The repeated use of illegal drugs or the inappropriate use of legal drugs to increase feelings of pleasure, reduce stress, and/or alter reality. [1]

Drugs of abuse: Illegal drugs (e.g., cocaine, marijuana, inhalants, PCP, methamphetamine) that are used repeatedly or legal drugs (e.g., alcohol) that are used inappropriately to increase feelings of pleasure, reduce stress, and/or alter reality.

E

Ecstasy (MDMA): A chemically modified amphetamine that has hallucinogenic as well as stimulant properties. [1]

Electrical message: A signal that begins in the axon terminal, is converted to a chemical message as it is carried across a synapse from one neuron to another, then is converted back to an electrical message when it arrives at a receptor.

Ethics: Refers to ways of understanding and examining moral issues that are shaped by individual, community, and societal values. Ethical reasoning involves a systematic process that generates acceptable, justifiable choices or options. [12,13]

Experiment: A test under controlled conditions that is made to demonstrate a known scientific truth or to test an unproven hypothesis. [5, adapted]

Explicit: Fully and clearly defined or expressed; leaving nothing implied. [5]

F

Frontal lobe: One of the four divisions of each hemisphere of the brain. The frontal lobe controls thought, speech, and smell.

G

GABA (gamma-aminobutyric acid): An amino acid, $C_4H_9NO_2$, that is not found in proteins, but occurs in the central nervous system and is associated with the transmission of nerve impulses. [5]

Genetic research: Genetic research is the scientific study of heredity. Parents pass on genetic information to their biological children through DNA. DNA is organized into genes and then chromosomes.

H

Hallucinogens: A diverse group of drugs that alter perceptions, thoughts, and feelings. Hallucinogenic drugs include lysergic acid diethylamide (LSD), mescaline, 3,4-methylenedioxymethamphetamine (MDMA or ecstasy), phencyclidine (PCP), and psilocybin ("magic mushrooms"). [1]

Heroin: An addictive opiate drug that produces a surge of euphoria followed by alternately wakeful and drowsy states and cloudy mental functioning. Overdoses of heroin can be fatal. [1, adapted]

Hormone: A chemical substance formed in glands in the body and carried in the blood to organs and tissues, where it influences function, structure, and behavior. [1]

Hypothalamus: The part of the brain that controls many bodily functions, including feeding, drinking, and the release of many hormones. [1]

Hypothesis: A tentative explanation for an observation, phenomenon, or scientific problem that can be tested by further investigation. [5]

Hypothesize: To make an educated guess that may explain a scientific problem; to speculate or theorize. [5]

I

Illicit: Not sanctioned by custom or law; unlawful. [5]

Implicit: (1) Implied or understood though not directly expressed: an implicit agreement not to raise the touchy subject. (2) Contained in the nature of something though not readily apparent. (3) Having no doubts or reservations; unquestioning: implicit trust. [5]

Infer: To conclude from evidence or premises. [5]

Inhalant: Any drug administered by breathing in its vapors. The chemicals in these vapors can produce feelings of pleasure for a short time. Inhalants commonly are organic solvents, such as glue and paint thinner, or anesthetic gases, such as ether and nitrous oxide. [1]

Institute for Laboratory Animal Research (ILAR): This private, nonprofit organization is responsible for developing and disseminating information on humane care and appropriate inclusion of animals. [3, adapted]

Institutional Animal Care and Use Committee (IACUC): Committee established to oversee and evaluate all aspects of a research institution's animal care and use program. [11, adapted]

In vitro: Literally "in glass," as in "in a test tube." A test that is performed in vitro is one that is done in glass or plastic vessels in the laboratory. [4, adapted]

In vivo: Literally "in the living organism," as opposed to in vitro (in the laboratory). A test that is done in vivo is one that is done in a living animal or human being. [4, adapted]

J

Justice: The fair and equal distribution of benefits and burdens on society, treating others fairly, and having the right to equal opportunity. [12,13]

L

Licit: Permitted by law; legal. [5]

Limbic system: A set of brain structures that generates our feelings, emotions, and motivations. It is also important in learning and memory. [1]

LSD (lysergic acid diethylamide): A hallucinogenic drug that acts on the serotonin receptor. [1]

M

Magnetic resonance imaging (MRI): A special radiology technique that uses magnetism, radio waves, and a computer to produce the images of internal structures of the body. [4, adapted]

Marijuana: A drug, usually smoked but can be eaten, that is made from the leaves of the cannabis plant. The main psychoactive ingredient is THC (delta-9-tetrahydrocannabinol). [1]

Methamphetamine: A commonly abused, potent stimulant drug that is part of a larger family of amphetamines. [1]

Model: A similar object or a reconstruction used to help visualize or explore something else that cannot be directly observed or experimented on (such as the living human body). [8, adapted]

Morals: Of or concerned with the judgment of "goodness" or "badness" of human action and character; conforming to standards of what is right and just in behavior; rules or habits or conduct with reference to standards of right and wrong. [5]

Motor cortex: The area of the cerebral cortex where impulses from the nerve centers to the muscles originate. [5]

Myelin: The fatty substance that covers and protects nerves.

N

Neuron (nerve cell): A unique type of cell found in the brain and body that is specialized to process and transmit information. [1]

Neuroscience: The study of the brain and other components of the nervous system, including the spinal cord and nerves. Neuroscience is also the study of diseases of the nervous system. [4, adapted]

Neurotransmitter: A chemical substance, such as dopamine, that transmits nerve impulses across a synapse. [5] The neurotransmitter acts like a "key" to open the receptor, which acts like a "lock." The two work together to deliver messages successfully.

Nicotine: The addictive drug in tobacco. Nicotine activates a specific type of acetylcholine receptor. [1]

Non-maleficence: The obligation to avoid doing intentional harm and working to prevent predictable harms from occurring to people or animals. [12,13]

Nucleus: A cluster or group of nerve cells that is dedicated to performing its own special function(s). Nuclei are found in all parts of the brain but are called cortical fields in the cerebral cortex. [1]

O

Observe: (1) To be or become aware of, especially through careful and directed attention; notice. (2) To watch attentively: observe a child's behavior. (3) To make a systematic or scientific observation of: observe the orbit of the Moon. [5]

Occipital lobe: One of the four divisions of each hemisphere of the brain. The occipital lobe controls the sense of sight.

Opiate: A medication or illegal drug that is either derived from the opium poppy, or that mimics the effect of an opiate (a synthetic opiate). Opiate drugs are narcotic sedatives that depress activity of the central nervous system, reduce pain, and induce sleep. Side effects may include oversedation, nausea, and constipation. Long-term use of opiates can produce addiction, and overuse can cause overdose and potentially death. [4]

Opinion: A personal belief or judgment that is not founded on proof or certainty. [7]

P _____

Paranoia: A psychotic disorder characterized by extreme, irrational distrust of others. [5]

Parietal lobe: One of the four divisions of each hemisphere of the brain. The parietal lobe controls sensory processes, such as touch and pain, attention, and language.

Population-based research: Population-based research studies a representative sample of a target population. The research results are then analyzed and applied to the entire target population.

Positron emission tomography (PET): A highly specialized imaging technique that uses short-lived radioactive substances to produce three-dimensional colored images of those substances functioning within the body. PET has been used primarily in cardiology, neurology, and oncology. [4]

Predict: To state, tell about, or make known in advance, especially on the basis of special knowledge. [5]

Principal investigator: In biomedical research, the person who directs a research project or program. The principal investigator (PI) usually writes and submits the grant application, oversees the scientific and technical aspects of the grant, and has responsibility for the management of the research. [4]

Profile: (1) A short description of a person's characteristics. (2) A formal summary or analysis of data, often in the form of a graph or table, representing distinctive features or characteristics: a psychological profile of a job applicant; a biochemical profile of blood. [5, adapted]

Protocol: The plan for a course of medical treatment or for a scientific experiment. [5]

R _____

Receptor: A large molecule that receives the message carried by the neurotransmitter. The receptor acts like a "lock," while the neurotransmitter acts like a "key." The two work together to deliver messages successfully.

Research: Scholarly or scientific investigation or inquiry. [5]

Research facility: Research facilities are places where pharmaceutical, biomedical, manufacturing, biotechnology, and other studies are conducted. An animal research facility is known as a vivarium and is carefully designed to provide an environment that ensures the care and maintenance of experimental animals.

Research subjects: Any organism (animal or plant) that is recruited for observation or study. When human beings are recruited as research subjects, they must agree to be subjects. When animals are included as research subjects, laws and guidelines ensure their ethical treatment. [5]

Reward: The process that reinforces behavior. It is mediated at least in part by the release of dopamine into the brain, which causes human subjects to feel a sense of pleasure. [1, adapted]

STUDENT GLOSSARY

S

Sanitize: Make less offensive or more acceptable by removing objectionable features: "sanitize a document before releasing it to the press"; "sanitize history"; "sanitize the language in a book." [7]

Scientific method: The principles and processes that guide scientific inquiry. The scientific method entails that an investigator: (1) state the problem or question, (2) collect information, (3) form a hypothesis, (4) test the hypothesis, (5) draw a conclusion, and (6) communicate results.

Seizure: Uncontrolled electrical activity in the brain, which may produce a physical convulsion, minor physical signs, thought disturbances, or a combination of symptoms. [4]

Serotonin: A neurotransmitter that regulates many functions, including mood, appetite, and sensory perception. [1]

Soma: The cell body of a neuron, where the nucleus is kept.

Spinal column: Series of vertebrae encasing the spine. The spinal column protects the spinal cord.

Spinal cord: The major column of nerve tissue that is connected to the brain and lies within the spinal column and from which the spinal nerves emerge. The spinal cord and the brain constitute the central nervous system (CNS). [4]

Sterile: (1) Not producing or incapable of producing offspring. (2) Free from live bacteria or other microorganisms: a sterile operating area; sterile instruments. [5]

Steroid: Artificial versions of the hormone testosterone, which regulates male sexual traits and muscle growth. [10]

Stimulants: A class of drugs that elevates mood, increases feelings of well-being, and increases energy and alertness. These drugs produce euphoria and are powerfully rewarding. Stimulants include cocaine methamphetamine, and methylphenidate (Ritalin). [1]

Synapse (a.k.a. synaptic cleft, space, or gap): The gap between neurons. Messages are carried by neurotransmitter from an axon terminal across the synaptic gap to a receptor.

T

Temporal lobe: One of the four divisions of each hemisphere of the brain. The temporal lobe controls memory, hearing, and language.

Thalamus: Located deep within the brain, the thalamus is the key relay station for sensory information flowing into the brain, filtering out important messages from the mass of signals entering the brain. [1]

THC: Delta-9-tetrahydrocannabinol; the main active ingredient in marijuana, which acts on the brain to produce its effects. [1]

Three Rs: Three important ethical alternatives scientists consider when including animals in research. The three Rs stand for:

Refine—Altering tests so that an animal's pain and/or distress are decreased to the absolute minimum as possible while keeping the data reliable and valid (pain management).

Replace—Using non-animals instead of animals as research subjects or including animals with less sensory perception if non-animal methods are not possible.

Reduce—Lowering the number of animals included in a specific research project or using newer statistical techniques that may give similar data quality.

Tobacco: A plant widely cultivated for its leaves, which are used primarily for smoking; the *tabacum* species is the major source of tobacco products. [1]

Tolerance: A condition in which higher doses of a drug are required to produce the same effect as during initial use; often leads to physical dependence. [1]

U _____

United States Department of Agriculture (USDA): The federal agency responsible for enforcement of the Animal Welfare Act.

V _____

Values: Express what is considered desirable, free from harm, and worthy in the common culture. Values include dignity, integrity, mutual respect, loyalty, friendship, fairness, and inclusiveness. [12,13]

Verify: To prove the truth by presentation of evidence or testimony; to substantiate. [5]

Veterinary Medical Officers (VMOS): Public health professionals who help oversee the effectiveness of the farm-to-the-table food safety system.

W _____

Withdrawal: Symptoms that occur after chronic use of a drug is reduced or stopped. [1]

GLOSSARY REFERENCES

1. National Institute on Drug Abuse. NIDA for Teens Glossary. *http://teens.drugabuse.gov/utilities/ glossary.asp#A*

2. Association for Assessment and Accreditation of Laboratory Animal Care International (AAALAC) International. *www.aaalac.org/about/index.cfm*

3. National Institutes of Health, Office of Extramural Research, Office of Lab and Animal Welfare. *http:// grants.nih.gov/grants/olaw/tutorial/glossary.htm*

4. MedicineNet.com. *www.medterms.com/script/ main/alphaidx.asp?p=a_dict*

5. The American Heritage Dictionary of the English Language, 4th ed. *http://dictionary.reference.com*

6. Webster's New Millennium Dictionary of English. Vol. 0.9.6. *http://dictionary.reference.com*

7. WordNet 2.0. *http://dictionary.reference.com*

8. Merriam-Webster's Medical Dictionary, 2002. *http://dictionary.reference.com*

9. Online Medical Dictionary. *http://dictionary. reference.com*

10. National Institute on Drug Abuse. 2000. The Brain's response to hallucinogens, inhalants, marijuana, methamphetamine, opiates, nicotine, steroids, stimulants [Series of 8 Brochures]. Rockville, MD: National Clearinghouse for Alcohol and Drug Information.

11. Institutional Animal Care and Use Committee (IACUC). General Information. *www.iacuc.org*

12. Reich, W.T., ed. 1995. *Encyclopedia of bioethics.* New York: Macmillan Publishing Company.

13. Roberts, L. W., and A. R. Dyer. 2004. *Ethics in mental health care.* Washington, DC: American Psychiatric Publishing.

When a definition was edited to better suit the needs of a middle school audience, the word "adapted" appears next to the reference number.

THIS IS YOUR BRAIN

Dear Parent or Guardian,

OVERVIEW

For our next unit in science, titled *This Is Your Brain*, we will explore the topic of drug abuse from a new perspective. Our students often hear the message that drugs are "bad" for them. In this unit, rather than attempting to scare students, we encourage students to become informed. Students will learn exactly *what happens* to the human brain when drugs are introduced, the damage done by drugs, and why biomedical research is so important in helping us understand the nature of drug abuse and addiction.

WHAT WE WILL BE DOING

First, students will learn how the human brain functions normally. Following this introduction to basic concepts in neuroscience, students will learn how drugs disrupt normal functioning of the brain and how the brain becomes dependent on drugs. As they learn these concepts, students will

1. gain solid instruction and practice in the scientific method;

2. learn about the importance of biomedical research in the prevention and treatment of drug addiction;

3. study and explore the field of ethics as it relates to biomedical research, a very relevant topic in today's fast-paced world of science discoveries; and

4. apply their understanding of ethical principles to the issue of using animals in drug abuse research.

HOMEWORK EXTENSIONS

For each lesson, interactive learning activities and homework extensions will serve to engage students and reinforce scientific concepts. Encourage your child to share his/her Project Portfolio with you on a daily basis and ask what he/she learned that day. You are also encouraged to visit the website resources provided with the program and try out the interactive computer game provided with the unit.

Please sign at the bottom of this letter indicating you have received and read this information and return it to school as soon as possible. Also, feel free to jot down any questions or comments you have about the unit. Don't hesitate to contact me at school. Thank you for your support.

_____

Teacher's Signature

Questions/Comments:

_____
_____
_____
_____
_____
_____
_____

_____

Parent's or Guardian's Signature

This project has been funded in whole or in part with federal funds from the National Institute on Drug Abuse, National Institutes of Health, Department of Health and Human Services, under Contract No. HHSN-271-2004-11124C.

TEACHER RESOURCES

Substance Abuse and Addiction

Books and Periodicals

Budney, A. J., and J. Wiley. 2001. Can marijuana use lead to marijuana dependence? In *Animal research and human health: Advancing human welfare through behavioral science*, ed. M. E. Carroll, and J. B. Overmier, 115–126. Washington, DC: American Psychological Association.

Cartwright, W. S. 1999. Costs of drug abuse to society. *The Journal of Mental Health Policy and Economics* 2: 133–134.

Friedman, D. P., and S. Rusche. 1999 *False messengers: How addictive drugs change the brain.* New York: Taylor and Francis.

Grabish, B. R. 1998. *Drugs and your brain.* New York: The Rosen Publishing Group.

Kandel, D. B., ed. 2002. *Stages and pathways of drug involvement: Examining the gateway hypothesis.* New York: Cambridge University Press.

Koob, G. 1997. The neurobiology of addiction: An overview. *Alcohol Health and Research World* 21 (2): 101–106.

Kosten ,T. R., T. F. Newton, et al., eds. 2011. *Cocaine and methamphetamine dependence: Advances in treatment.* Washington, DC: American Psychiatric Press.

Martin, K. R. 2003. Youths' opportunities to experiment influence later use of illegal drugs. [Electronic version]. *NIDA Notes* 17 (5).

Roberts, T. G., G. P. Fournet, and E. Penland. 1995. A comparison of the attitudes toward alcohol and drug use and school support by grade level, gender, and ethnicity. *Journal of Alcohol and Drug Addiction* 40 (2): 112–127.

Yudofsky S., and R. E. Hales. 2010. *Essentials of neuropsychiatry and behavioral neuroscience.* Washington, DC: American Psychiatric Press.

Websites

The Centers for Disease Control and Prevention. Alcohol. *www.cdc.gov/alcohol/index.htm*

The Centers for Disease Control and Prevention. Tobacco. *www.cdc.gov/tobacco/index.htm*

National Household Survey on Drug Abuse. Academic performance and youth substance use. *http://oas.samhsa.gov/2k2/academics/academics.htm*

National Institute on Drug Abuse (NIDA). The brain and addiction. *http://teens.drugabuse.gov/facts/facts_brain1.asp*

National Institute on Drug Abuse (NIDA). InfoFacts. *www.drugabuse.gov/NIDAHome.html*

National Institute on Drug Abuse (NIDA). Information Sheets on common drugs such as tobacco, alcohol, marijuana, cocaine, etc. These informational sheets can be found in the NIDA Info Facts section of the NIDA website. *www.nida.nih.gov/infofax/InfofaxIndex.html*

National Institute on Drug Abuse (NIDA). NIDA Notes

covers drug abuse research in the areas of treatment and prevention, epidemiology, neuroscience, behavioral science, health services, and AIDS. *www. nida.nih.gov/NIDA_Notes/NNIndex.html*

National Institute on Drug Abuse (NIDA). The science behind drug abuse: The brain and addiction. *http:// teens.drugabuse.gov/facts/facts_brain1.asp*

National Institute on Drug Abuse (NIDA). The science behind drug abuse: Mind over matter, Teacher's guide. *http://teens.drugabuse.gov/mom/tg_intro.asp*

National Institute on Drug Abuse (NIDA). The science of drug abuse and addiction: NIDA homepage. *http:// www.drugabuse.gov/NIDAHome.html*

RAND Corporation. RAND Corporation is a nonprofit research organization that conducts research to address issues that affect the world. *www.rand.org/ research_areas/substance_abuse*

Substance Abuse and Mental Health Services Administration (SAMHSA). National Clearinghouse for Alcohol and Drug Information (NCADI). List of posters to order: *https://store.health.org/catalog/ results.aspx?h=publications&topic=124*

Substance Abuse and Mental Health Services Administration (SAMHSA). National Clearinghouse for Alcohol and Drug Information (NCADI). List of publications to order: *http://store.health.org/ catalog/results.aspx?h=audiences&topic=10*

Substance Abuse and Mental Health Services Administration (SAMHSA). National Clearinghouse for Alcohol and Drug Information (NCADI). List of publications and materials related to community coalitions. *www.health.org/features/community*

Substance Abuse and Mental Health Services Administration (SAMHSA). National Clearinghouse for Alcohol and Drug Information (NCADI). List of publications and materials related to educators. *www.health.org/features/school*

Substance Abuse and Mental Health Services Administration (SAMHSA). National Clearinghouse for Alcohol and Drug Information (NCADI). List of publications and materials related to parents and caregivers. *www.health.org/features/family*

Substance Abuse and Mental Health Services Administration (SAMHSA). National Clearinghouse for Alcohol and Drug Information (NCADI). List of publications and materials related to teens and children. *www.health.org/features/youth*

University of Michigan. Monitoring the future: A continuing study of American youth. *www. monitoringthefuture.org*

U.S. Department of Health and Human Services and SAMHSA's National Clearinghouse for Alcohol and Drug Information. *School. www.health.org/ features/school*

National Institute on Drug Abuse (NIDA) offers Slide Presentations that can be downloaded and used in the classroom. *www.drugabuse.gov/pubs/ Teaching*

Topics of slide shows include:

> The Brain & the Actions of Cocaine, Opiates, and Marijuana
>
> The Neurobiology of Drug Addiction
>
> Understanding Drug Abuse and Addiction: What Science Says
>
> The Neurobiology of Ecstasy (MDMA)
>
> Bringing the Power of Science to Bear on Drug Abuse and Addiction

Animals in Research

Books and Periodicals

Bishop, L. J., and A. L. Nolen. 2001. Animals in research and education: Ethical issues. *Kennedy Institute of Ethics Journal* 11: 91–112.

Botting, J. H., and A. Morrison. 1997. Animal research

is vital to medicine. *Scientific American* 276 (2): 83–85.

Carroll, M. E., and J. B. Overmier. 2001. *Animal research and human health*. Washington, DC: American Psychological Association.

DeGrazia, D. 1999. Animal ethics around the turn of the twenty-first century. *Journal of Agricultural and Environmental Ethics* 11: 111–129.

Dennis, J. U. 1997. Morally relevant differences between animals and human beings justify the use of animals in biomedical research. *Journal of the American Veterinary Medical Association* 210 (5): 612–618.

Fitzpatrick, A. 2003. Ethics and animal research. *Journal of Laboratory and Clinical Medicine* 141 (2): 89–90.

Gluck, J. P., and J. Bell. 2003. Ethical issues in the use of animals in biomedical and psychopharmacological research. *Psychopharmacology* 171: 6–12.

Gluck, J. P., T. DiPasquale, and F. B. Orlans. 2002. *Applied ethics in animal research*. West Lafayette, IN: Purdue University.

Join Hands. 1999. *Alternative research methods, refinement, reduction, replacement of animals needed in scientific research*. 1-800-933-8288.

Mukerjee, M. 1997. Trends in animal research. *Scientific American* 276 (2): 86–94.

National Research Council. 1996. *Guide for the care and use of laboratory animals*. Washington, DC: National Academies Press.

Page, G. G. 2004. The importance of animal research in nursing science. *Nursing Outlook* 52 (4): 102–107.

Pitts, M. ed. 2003. *Institutional animal care and use committee guidebook*. Collingdale, PA: DIANE Publishing.

Rowen, A. N. 1997. The benefits and ethics of animal research. *Scientific American* 276 (2): 79.

Tabakoff, B., and P. L. Hoffman. 2000. Animal models. Part 1: Behavior and physiology. *Alcohol Research and Health* 24: 77–144.

Witek-Janusek, L. 2004. Commentary on the importance of animal research in nursing science, *Nursing Outlook* 52 (4): 108–110.

Websites

American Psychological Association (APA). Committee on animal research and ethics. *www.apa.org/science/resethicsCARE.html*

American Psychological Association (APA). Guidelines for ethical conduct in the care and use of animals. *www.apa.org/science/anguide.html*

Americans for Medical Progress. Animal research. Timeline of animal contributions to medical treatment and technique development features breakthroughs from pre-1900s through the 1990s. *www.amprogress.org/Issues?IssuesList.cfm?c=10*

Animal Welfare Information Center. Questions and answers about the animal welfare act and its regulations for biomedical research institutions. *www.nalusda.gov/awic/legislat/regsqa.htm*

Animal Welfare Institute. Animals in Laboratories. *www.awionline.org/lab_animals/index.htm*

Center for Alternatives to Animal Testing (CAAT). Homepage. *http://caat.jhsph.edu*

Committee on Animal Research and Ethics in the Care and Use of Animals. Guidelines for ethical conduct in the care and use of animals. *www.apa.org/science/anguide.html*

Foundation for Biomedical Research. Nobel Prizes for medical and physiological breakthroughs involving animal research are noted with reference to specific animals involved in each discovery. *www.fbresearch.org/education/nobels.htm*

The Human Society of the United States. Animals in research. *www.hsus.org/animals_in_research*

Johns Hopkins Bloomberg School of Public Health. Altweb: The global clearinghouse for information on alternatives to animal testing. *http://altweb.jhsph.edu*

Johns Hopkins University. Center for alternatives to animal testing. *http://caat.jhsph.edu*

Kids 4 Research. Responsible laboratory animal care and use standards are provided to teachers and students along with information on benefits of animal research to animals, humans, and the environment. *www.kids4research.org*

Koob, G. F. 1995. Animal models of drug addiction. Psychopharmacology: The Fourth Generation of Progress. 4th ed., Raven Press. *www.acnp.org/g4/GN401000072/CHO72.html*

Massachusetts Society for Medical Research. Promotes biomedical and biological research, including the humane care and use of animals for the improved health and well-being of people, animals, and the environment. *www.msmr.org*

Michigan Society for Medical Research. Education materials. *www.mismr.org/educational*

Minnesota Branch of the American Association for Laboratory Animal Science. Written at a middle school level, this site discusses the definition of animal models, justification for their use in medical research, and the development of standards for care and use of laboratory animals. A link to a diagram of the research process is also provided. *www.ahc.umn.edu/rar/MNAALAS/Models.html*

National Agricultural Library. Questions and answers about the animal welfare act and its regulations for biomedical research institutions. *www.nalusda.gov/awic/legislat/regsqa.htm*

The Research Defence Society. Understanding animal research in medicine: Animal research facts. *www.rds-online.org.uk/pages/page.asp?i_ToolbarID=2&i_PageID=48*

The Research Defence Society. Understanding animal research in medicine: Medical milestones. *www.rds-online.org.uk/pages/page.asp?i_ToolbarID=3&i_PageID=3*

The Research Defence Society. Understanding animal research in medicine: Animal welfare. *www.rds-online.org.uk/pages/page.asp?i_ToolbarID=4&i_PageID=4*

United States Department of Agriculture. Animal welfare act. *www.nal.usda.gov/awic/legislat/awa.htm*

University of Minnesota: Academic Health Center. Animal models of disease: A diagram of the research process. *www.ahc.umn.edu/rar/MNAALAS/Models.html*

Ethics

Books and Periodicals

Bebeau, M. J., J. R. Rest, and D. Narvaez. 1999. Beyond the promise: A perspective on research in moral education. *Educational Researchers* 28 (4): 18–26.

Benninga, J. S., M. W. Berkowitz, P. Kuehn, and K. Smith. 2003. The relationships of character education and academic achievement in elementary schools. *Journal of Research in Character Education* 1 (1): 17–30.

Brooks, B. D., and M. E. Kann. 1993. The school's role in weaving values back into the fabric of society. *Education Digest* 58 (8): 67–71.

Character Education Partnership. (2002). *Practices of teacher educators committed to characters. Examples from teacher education programs emphasizing character development.* Washington, DC: Character Education Partnership.

Goleman, D. 1995. *Emotional intelligence: Why it can matter more than IQ.* New York: Bantam.

Ryan, K., and K. E. Bohin. 1999. *Building character in schools. Practical ways to bring moral instruction to life.* San Francisco: Jossey-Bass.

Shapiro, D. A. 1999. Teaching ethics from the inside-out:

Some strategies for developing moral reasoning skills in middle-school students. Paper presented to the Seattle Pacific University Conference on the Social and Moral Fabric of School Life, Edmonds, WA.

Websites

The Online Ethics Center for Engineering and Science at Case Western Reserve University. Ethics in the science classroom: Introduction. *http://onlineethics. org/education/precollege/scienceclass/introethics. aspx*

Science Education

Books and Periodicals

American Association for the Advancement of Science. 1990. *Science for all Americans: Project 2061*. New York: Oxford University Press.

American Association for the Advancement of Science. 1993. *Benchmarks for science literacy*. New York: Oxford University Press.

American Association for the Advancement of Science. 1997. *Resources for science literacy: Professional development*. New York: Oxford University Press.

Craven, J. A. III, and T. Hogan. 2001. Assessing student participation in the classroom. *Science Scope* 25 (1): 36–40.

Herzog, H. A. 1990. Discussing animal rights and animal research in the classroom. *Teaching of Psychology* 17 (2): 90–94.

Johnson, D. W., and R. T. Johnson. 1996. *Meaningful and manageable assessment through cooperative learning*. Edina, MN: Interaction Book Company.

Jorgenson, O., J. Cleveland, and R. Vanosdall. 2004. *Doing good science in middle school: A practical guide to inquiry-based instruction*. Arlington, VA: NSTA Press.

Kelly, A. E., and R. A. Lesh. 2000. *Handbook of research design in mathematics and science education*. Mahwah, NJ: Lawrence Erlbaum Association.

National Research Council. 1996. *National science education standards*. Washington, DC: National Academies Press.

North Carolina Association for Biomedical Research. 2001. What's the point of bioscience research? (919) 785-1304.

Rakow, S. J., ed. 2000. *NSTA pathways to the science education standards: Middle school edition*. Arlington, VA: NSTA Press.

Stevenson, C. 1992. *Teaching ten to fourteen year olds*. New York: Longman Publishers.

Sylwester, R. 1996. *A celebration of neurons: An educator's guide to the human brain*. Alexandria, VA: Association for Supervision and Curriculum Development.

Websites

Centre for Neuro Skills. Brain functions and map. *www. neuroskills.com/brain-injury/brain-function.php*

How Stuff Works. *http://science.howstuffworks.com/ brain.htm*

The John Hopkins Bayview Medical Center. Links to current articles about topics that are new in healthcare. *www.jhbmc.jhu.edu/healthcarenews/ index.html*

Kids 4 Research. Grades 7–12 educational materials. www.kids4research.org/712.html and Grades K–6 educational materials. *www.kids4research.org/ k6.html*

National Institute of Environmental Health Sciences. *www.niehs.nih.gov*

Neuroscience for Kids. *http://faculty.washington.edu/ chudler/neurok.html*

PARENT RESOURCES

Substance Abuse and Addiction

Books and Periodicals

Emmett, D., and G. Nice. 1996. *Understanding drugs: A handbook for parents, teachers and other professionals*. London: Jessica Kingsley.

Friedman, D. P., and S. Rusche. 1999. *False messengers: How addictive drugs change the brain*. New York: Taylor and Francis.

Grabish, B. R. 1998. *Drugs and your brain.* New York: The Rosen Publishing Group.

Martin, K. R. 2003. Youths' opportunities to experiment influence later use of illegal drugs. [Electronic version] *NIDA Notes* 17 (5).

Yudofsky S. C., and R. E. Hales. 2010. *Essentials of neuropsychiatry and behavioral neurosciences.* Washington DC: American Psychiatric Press.

Websites

American Council for Drug Education. Facts for parents. *www.acde.org/parent/Default.htm*

The Centers for Disease Control and Prevention. Alcohol. *www.cdc.gov/alcohol/index.htm*

The Centers for Disease Control and Prevention. Tobacco. *www.cdc.gov/tobacco/index.htm*

FDC Educational Services. Resources for parents and guardians. *www.fcd.org/content/resources/parents.asp*

National Center for Chronic Disease Prevention and Health Promotion. Tobacco information and prevention source (TIPS): Tips 4 youth. *www.cdc.gov/tobacco/tips4youth.htm*

National Families in Action. A guide to drugs and the brain. *www.nationalfamilies.org/brain/index.html*

National Household Survey on Drug Abuse. Academic performance and youth substance use. *http://oas.samhsa.gov/2k2/academics/academics.htm*

National Institute on Drug Abuse (NIDA). The brain and addiction. *http://teens.drugabuse.gov/facts/facts_brain1.asp*

National Institute on Drug Abuse (NIDA). Parents and teachers. *www.nida.nih.gov/parent-teacher.html*

National Institute on Drug Abuse (NIDA). The science behind drug abuse: Facts on drugs. *http://teens.drugabuse.gov/facts/facts_brain1.asp*

National Institute on Drug Abuse (NIDA). The science of drug abuse and addiction. *www.drugabuse.gov/NIDAHome.html*

Substance Abuse and Mental Health Services Administration (SAMHSA). List of publications and materials related to community coalitions. *www.health.org/features/community*

Substance Abuse and Mental Health Services Administration (SAMHSA). List of publications and materials related to parents and caregivers. *www.health.org/features/family*

Substance Abuse and Mental Health Services Administration (SAMHSA). List of publications and materials related to teens and children. *www.health.org/features/youth*

Substance Abuse and Mental Health Services Administration (SAMHSA). List of publications and materials related to the workplace. *www.health.org/features/workplace*

University of Michigan. Monitoring the future: A continuing study of American youth. *www.monitoringthefuture.org*

USA Today. 1999. Need to know. *www.steponline.com/need/seco_need.asp*

Medical Research With Animals

Books and Periodicals

Botting, J. H., and A. Morrison. 1997. Animal research is vital to medicine. *Scientific American* 276 (2): 83–85.

Carroll, M. E., and J. B. Overmier. 2001. *Animal research and human health*. Washington, DC: American Psychological Association.

Join Hands. 1999. *Alternative research methods, refinement, reduction, replacement of animals needed in scientific research*. 1-800-933-8288.

Mukerjee, M. 1997. Trends in animal research. *Scientific American* 276 (2): 86–94.

National Research Council. 1996. *Guide for the care and use of laboratory animals*. Washington, DC: National Academies Press.

Pitts, M., ed. 2003. *Institutional animal care and use committee guidebook*. Collingdale, PA: DIANE Publishing.

Rowen, A. N. 1997. The benefits and ethics of animal research. *Scientific American* 276 (2): 79.

Websites

American Association for Laboratory Animal Science. Providing information to students, teachers, and parents on responsible laboratory animal care and use in biomedical research, testing, and education: Information for parents: *www.kids4research.org/parents.html*

American Psychological Association. Committee on animal research and ethics. *www.apa.org/science/resethicsCARE.html*

American Psychological Association. Guidelines for ethical conduct in the care and use of animals. *www.apa.org/science/anguide.html*

Americans for Medical Progress: Animal research. Timeline of animal contributions to medical treatment and technique development features breakthroughs from pre-1900s through the 1990s. *www.amprogress.org/Issues?IssuesList.cfm?c=10*

Animal Welfare Information Center. Questions and answers about the animal welfare act and its regulations for biomedical research institutions. *www.nalusda.gov/awic/legislat/regsqa.htm*

Animal Welfare Institute. Animals in Laboratories. *www.awionline.org/lab_animals/index.htm*

Center for Alternatives to Animal Testing (CAAT). *http://caat.jhsph.edu*

Committee on Animal Research and Ethics in the Care and Use of Animals. Guidelines for ethical conduct in the care and use of animals. *www.apa. org/science/anguide.html*

Foundation for Biomedical Research. Nobel Prizes for medical and physiological breakthroughs involving animal research are noted with reference to specific animals involved in each discovery. *www.fbresearch. org/education/nobels.htm*

The Human Society of the United States. Animals in research. *www.hsus.org/animals_in_research*

Kids 4 Research. Responsible laboratory animal care and use standards are provided to teachers and students along with information on benefits of animal research to animals, humans, and the environment. Site focus includes biomedical/ biological research and education. *www. kids4research.org*

Minnesota Branch of the American Association for Laboratory Animal Science Inc. Written at a middle school level, this site discusses the definition of animal models, justification for their use in medical research, and the development of standards for care and use of laboratory animals. A link to a diagram of the research process is also provided. *www.ahc.umn.edu/rar/MNAALAS/Models.html*

National Agricultural Library. Questions and answers about the animal welfare act and its regulations for biomedical research institutions. *www.nalusda. gov/awic/legislat/regsqa.htm*

United States Department of Agriculture. Animal welfare act. *www.nal.usda.gov/awic/legislat/awa. htm*

Ethics

Books and Periodicals

Benninga, J. S., M. W. Berkowitz, P. Kuehn, and K. Smith. 2003. The relationships of character education and academic achievement in elementary schools. *Journal of Research in Character Education* 1 (1): 17–30.

Goleman, D. 1995. *Emotional intelligence: Why it can matter more than IQ.* New York: Bantam.

Roberts L. W., and A. Dyer. 2004 *Ethics in mental health care.* Washington, DC: American Psychiatric Press.

Ryan, K., and K. E. Bohin. 1999. *Building character in schools. Practical ways to bring moral instruction to life.* San Francisco: Jossey-Bass.

Websites

The Online Ethics Center for Engineering and Science at Case Western Reserve University. Ethics in the science classroom: Introduction. *http://onlineethics.org/education/precollege/ scienceclass.asx*

The President's Council on Bioethics. Advising the President on ethical issues related to advances in biomedical science and technology. *www. bioethics.gov*

Science Education

Books and Periodicals

American Association for the Advancement of Science. 1993. *Benchmarks for science literacy.* New York: Oxford University Press.

Kramer, S. P. 1987. *How to think like a scientist:*

PARENT RESOURCES

Answering questions by the scientific method. New York: HarperCollins Children's Books.

National Research Council. 1996. *National science education standards.* Washington, DC: National Academies Press.

North Carolina Association for Biomedical Research. 2001. What's the point of bioscience research? (919) 785-1304.

Websites

Centre for Neuro Skills. Brain functions and map. *www.neuroskills.com/brain-injury/brain-function.php*

How Stuff Works. *http://science.howstuffworks.com/brain.htm*

The John Hopkins Bayview Medical Center. Links to current articles about topics that are new in healthcare. *www.jhbmc.jhu.edu/healthcarenews/index.html*

Kids 4 Research. Grades 7–12 educational materials. *www.kids4research.org/712.html* and Grades K–6 educational materials. *www.kids4research.org/k6.html*

National Institute of Environmental Health Sciences. *www.niehs.nih.gov*

Neuroscience for Kids. *http://faculty.washington.edu/chudler/neurok.html*

Office of Disease Prevention and Health Promotion, U.S. Department of Health and Human Services. *www.healthfinder.gov/kids/parents.htm*

University of Washington. Neuroscience for kids: Explore the brain and spinal cord. *http://faculty.washington.edu/chudler/introb.html#drug*

U.S. Food and Drug Administration. *www.fda.gov*

STUDENT RESOURCES

Substance Abuse and Addiction

Books and Periodicals

Friedman, D. P., and S. Rusche. 1999. *False messengers: How addictive drugs change the brain.* New York: Taylor and Francis.

Grabish, B. R. 1998. *Drugs and your brain.* New York: The Rosen Publishing Group.

Sparks, B., ed. 1998. *Go ask Alice.* New York: Simon and Schuster.

Websites

American Council for Drug Education. Facts for youth. *www.acde.org/youth/Default.htm*

The Centers for Disease Control and Prevention. Alcohol. *www.cdc.gov/alcohol/index.htm*

The Centers for Disease Control and Prevention. Tobacco. *www.cdc.gov/tobacco/index.htm*

FDC Educational Services. Resources for middle and high school students. *www.fcd.org/content/resources/students.asp*

National Center for Chronic Disease Prevention and Health Promotion. Tobacco information and prevention source (TIPS): The surgeon general's report for kids about smoking. *www.cdc.gov/tobacco/sgr/sgr4kids/sgrmenu.htm*

National Center for Chronic Disease Prevention and Health Promotion. Tobacco information and prevention source (TIPS): Tips 4 youth. *www.cdc.gov/tobacco/tips4youth.htm*

National Families in Action. A guide to drugs and the brain. *www.nationalfamilies.org/brain/index.html*

National Institute on Drug Abuse (NIDA). The brain and addiction. *http://teens.drugabuse.gov/facts/facts_brain1.asp*

National Institute on Drug Abuse (NIDA). Have fun and learn. *http://teens.drugabuse.gov/havefun/index.asp*

National Institute on Drug Abuse (NIDA). InfoFacts. *www.drugabuse.gov/NIDAHome.html*

National Institute on Drug Abuse (NIDA). Information Sheets on common drugs such as tobacco, alcohol, marijuana, and cocaine. *www.nida.nih.gov/infofax/InfofaxIndex.html*

National Institute on Drug Abuse (NIDA). The science behind drug abuse. *http://teens.drugabuse.gov/index.asp*

National Institute on Drug Abuse (NIDA). The science of drug abuse and addiction. *www.drugabuse.gov/NIDAHome.html*

National Institute on Drug Abuse (NIDA). The science behind drug abuse: The brain and addiction. *http://teens.drugabuse.gov/facts/facts_brain1.asp*

Substance Abuse and Mental Health Services Administration (SAMHSA). National Clearinghouse for Alcohol and Drug Information (NCADI). List of publications and materials related to teens and children. *www.health.org/features/youth*

U.S. Department of Health and Human Services and SAMHSA's National Clearinghouse for Alcohol and Drug Information. Youth. *www.health.org/features/youth*

Animals in Research

Books and Periodicals

Botting, J. H., and A. Morrison. 1997. Animal research is vital to medicine. *Scientific American* 276 (2): 83–85.

Mukerjee, M. 1997. Trends in animal research. *Scientific American* 276 (2): 86–94.

Rowen, A. N. 1997. The benefits and ethics of animal research. *Scientific American* 276 (2): 79.

Websites

Americans for Medical Progress: Animal research. Timeline of animal contributions to medical treatment and technique development features breakthroughs from pre-1900s through the 1990s. *www.amprogress.org/Issues?IssuesList.cfm?c=10*

Animal Welfare Institute. Animals in Laboratories. *www.awionline.org/lab_animals/index.htm*

Foundation for Biomedical Research. Nobel Prizes for medical and physiological breakthroughs involving animal research are noted with reference to specific animals involved in each discovery. *www.fbresearch.org/education/nobels.htm*

The Human Society of the United States. Animals in research. *www.hsus.org/animals_in_research*

Kids 4 Research. Information about responsible laboratory animal care and use in biomedical research, testing, and education. *www.kids4research.org*

Minnesota Branch of the American Association for Laboratory Animal Science Inc. Written at a middle school level, this site discusses the definition of animal models, justification for their use in medical research, and the development of standards for care and use of laboratory animals. A link to a diagram of the research process is also provided. *www.ahc.umn.edu/rar/MNAALAS/Models.html*

Science Education

Books and Periodicals

Kramer, S. P. 1987. *How to think like a scientist: Answering questions by the scientific method.* New York: HarperCollins Children's Books.

Websites

Centre for Neuro Skills. Brain functions and map. *www.neuroskills.com/brain-injury/brain-function.php*

How Stuff Works. *http://science.howstuffworks.com/brain.htm*

Kids 4 Research. *www.kids4research.org*

National Institute of Environmental Health Sciences (NIEHS), National Institutes of Health. NIEHS kids pages. *www.niehs.nih.gov/kids/home.htm*

Neuroscience for Kids. *http://faculty.washington.edu/chudler/neurok.html*

Neuroscience for Kids. Explore the nervous system. *http://faculty.washington.edu/chudler/introb.html*

Office of Disease Prevention and Health Promotion, U.S. Department of Health and Human Services.

Healthfinder kids. *http://www.healthfinder.gov/kids/default.htm*

University of Washington. Neuroscience for kids: Explore the brain and spinal cord. *http://faculty.washington.edu/chudler/introb.html#drug*

U.S. Food and Drug Administration. Kids home page. *http://www.fda.gov/oc/opacom/kids/default.htm*

UNIT REFERENCES

About, Inc. Mental health resources: Predisposition for drug abuse even before it starts. *http://mentalhealth. about.com/library/sci/1001/bldrugabuse1001.htm*

American Association for Laboratory Animal Science. Animal roles in medical discoveries. *www.aalas.org*

American Association for Laboratory Animal Science. The use of animals in biomedical research: Animal roles in medical discoveries. *www.aalas.org*

American Association for Laboratory Animal Science. The use of animals in biomedical research: Improving human and animal health. *www.aalas.org*

American Psychological Association. 2003. *Guidelines for the use of animals in behavioral projects in schools (K–12)*. Washington, DC: American Psychological Association.

American Psychological Association. 2005. APA guidelines for ethical conduct in the care and use of animals. *www.apa.org/science/anguide.html*

Americans for Medical Progress. AMP fact sheets: Why animals in research? Three facts about animal research in medicine. *www.amprogress.org/Issues/IssuesList.cfm?c=91*

Americans for Medical Progress. 2004. Animal welfare regulation: An overview. *www.amprogress.org/Issues/IssuesList.cfm?c=11*

Americans for Medical Progress. 2005. Benefits to animals. *www.amprogress.org/Issues/IssuesList.cfm?c=12*

Animal Welfare Information Center. Welfare act and its regulations for biomedical research institutions. *www.nalusda.gov/awic/legislat/regsqa.htm*

Beauchamp, T. L. 2001. *Principles of biomedical ethics.* New York: Oxford University Press.

Brooks, B. D., and M. E. Kann. 1993. The school's role in weaving values back into the fabric of society. *Education Digest* 58: 67–71.

BSCS and Videodiscovery. 2000. The brain: Understanding neurobiology through the study of addiction. *http://science.education.nih.gov/supplements/nih2/addiction/default.htm*

Campbell, T. Central Nervous System Disorders. PowerPoint Presentation of the Central Nervous System.

Cartwright, W. S. 1999. Costs of drug abuse to society. *The Journal of Mental Health Policy and Economics* 2: 133–134.

Center for Science, Mathematics, and Engineering Education (CSMEE). 1997. *Introducing the national science education standards* [Booklet]. Washington, DC: The National Academies Press.

Committee on Animal Research and Ethics (CARE). Research with animals in psychology. *www.apa.org/science/animal2.html*

Crowley, T. J., M. J. Macdonald, E. A. Whitmore, and S. K. Mikulich. 1998. Cannabis dependence, withdrawal, and reinforcing effects among adolescents with conduct symptoms and substance use disorders. *Drug and Alcohol Dependence* 50 (1): 27–37.

Donohue, D. Animal welfare act. Unpublished raw data.

Drug Enforcement Administration. Ecstasy use: The known dangers. *http://www.justice.gov/dea/pubs/abuse/drug_data_sheets/Ecstacy.pdf*

Foundation for Biomedical Research. Frequently asked questions about animal research. *www.fbresearch. org/animal-research-faq.htm*

Foundation for Biomedical Research. 2005. Nobel prizes: The payoff from animal research. *www.fbresearch. org/education/nobels.htm*

Freudenrich, C. How your brain works. *http://science. howstuffworks.com/brain.html*

Gluck, J. P., T. DiPasquale, and B. F. Orlans. 2002. *Applied ethics in animal research.* West Lafayette, IN: Purdue University Press.

Gordis, E. 2000. Animal models in alcohol research. *Alcohol Research & Health* 24 (2): 75–76.

Healthy people 2010: Substance abuse objectives. PowerPoint Presentation of Substance Abuse.

Herzog, H. A. 1990. Discussing animal rights and animal research in the classroom. *Teaching of Psychology* 17: 90–94.

Johnston, L. D., P. M. O'Malley, and J. G. Bachman. 2003. *Monitoring the future national results on adolescent drug use: Overview of key findings, 2002.* Bethesda, MD: National Institute on Drug Abuse.

Join Hands. 1999. Alternative research methods, refinement, reduction, replacement of animals needed in scientific research. 1-800-933-8288.

Kandel, D. B. 2002. *Stages and pathways of drug involvement: Examining the gateway hypothesis.* New York: Cambridge University Press.

Khantzian, E. J. 1987. A clinical perspective of the cause-consequence controversy in alcoholic and addictive suffering. *Journal of the American Academy of Psychoanalysis* 15: 521–537.

Koob, G. F. 2000. Animal models of drug addiction. In *Psychopharmacology: The Fourth Generation of Progress,* 4th ed., eds. F. E. Bloom and D. J. Kupfer, 1193–1204. New York: Raven Press.

Kukula, K. Amazing facts. Scholastic. *http://teacher. scholastic.com/scholasticnews/ indepth/headsup/ brain/index.asp?article=brain_facts*

Mathias, R., and N. Swan. 1995. Using animals to study mechanisms and effects of drugs. *NIDA Notes* (Nov/ Dec): 10.

Michigan Society for Medical Research. Fact vs. myth. *www.mismr.org/educational/factmyth.html*

Michigan Society for Medical Research. Fast facts about biomedical research. *www.mismr.org/educational/ fast_facts.pdf*

Michigan Society for Medical Research. Questions about biomedical research. *www.mismr.org/educational/ biomedres.html*

Michigan Society for Medical Research. Roles animals have played in biomedical research. *www.mismr.org/ educational/animal_roles.pdf*

Miller, N. S., and R. J. Goldsmith. 2001. Craving for alcohol and drugs in animals and humans: Biology and behavior. *Journal of Addictive Diseases* 20 87–104.

Minnesota Branch of the American Association for Laboratory Animal Science. 2000. Animals in science. *www.ahc.umn.edu/rar/MNAALAS/?Models. html*

Minnesota Branch of the American Association for Laboratory Animal Science Inc. 2000. Experimentation and the scientific method. *www. ahc.umn.edu/rar/MNAALAS/Method.html*

National Academy for Curriculum Leadership. SCI Center at BCS. 2002. *BSCS 5E Instructional Model.* Colorado Springs, CO: Biological Sciences Curriculum Study.

National Center for Education Statistics. 2002. Common core of data surveys. *http://nces.ed.gov/ccd*

National Center for Education Statistics. 1999. Parent survey of the national household education surveys program. *http://nces.ed.gov/pubs2001/homeschool*

National Institute on Alcohol Abuse and Alcoholism. 1994. Animal models in alcohol research (April): *Alcohol Alert Publication* No.24 PH350.

National Institute on Drug Abuse. The brain and addiction. *http://teens.drugabuse.gov/parents/parents_brain1.asp*

National Institute on Drug Abuse. The brain: Understanding neurobiology through the study of addiction. *http://science-education.nih.gov/supplements/nih2/addiction/ default.htm*

National Institute on Drug Abuse. Marijuana: Facts for teens. *www.nida.nih.gov*

National Institute on Drug Abuse. Mind over matter: The brain's response to opiates, nicotine, marijuana, methamphetamine, steroids, stimulants, inhalants, and hallucinogens. *http://teens.drugabuse.gov/mom/index.asp*

National Institute on Drug Abuse. NIDA InfoFacts: Cigarettes and other nicotine products. *www.nida.nih.gov/Infofacts/Tobacco.html*

National Institute on Drug Abuse. 1995. NIDA InfoFacts: Cost to society. *www.drugabuse.gov/Infofax/costs.html*

National Institute on Drug Abuse. NIDA InfoFacts: Crack and cocaine. *www.drugabuse.gov/Infofacts/cocaine.html*

National Institute on Drug Abuse. NIDA InfoFacts: Understanding drug abuse and addiction. *www.drugabuse.gov/Infofax/understand.html*

National Institute on Drug Abuse. NIDA InfoFacts: Workplace trends. *www.drugabuse.gov/Infofacts/workplace.html*

National Institute on Drug Abuse. Scholastic-Heads up: Real news about drugs and your body: Year 1 compilation. *www.drugabuse.gov/scholastic.html*

National Institute on Drug Abuse. 1998. The sixth triennial report to congress: Drug abuse and addiction research: 25 years of discovery to advance the health of the public. *www.drugabuse.gov/STRC/STRCindex.html*

National Institute on Drug Abuse. Understanding drug abuse and addiction—What science says. PowerPoint Presentation of the Science Behind Addiction.

National Institute on Out-of-School Time. 2004. Making the case: A fact sheet on children and youth in out-of-school time. *www.niost.org*

National Research Council. 1996. *Guide for the care and use of laboratory animals*. Washington, DC: National Academies Press.

National Research Council. 1996. *National science education standards*. Washington, DC: National Academies Press.

North Carolina Association for Biomedical Research. How many and what kinds of animals are used in biomedical research? *www.ncabr.org/biomed/FAQ_animal/faq_animal_13.html*

North Carolina Association for Biomedical Research. 1994. *What's the point of bioscience research?* Raleigh, NC: North Carolina Association for Biomedical Research.

Office of Applied Studies. 2000. *Summary of findings from the 1999 National Household Survey on Drug Abuse* (DHHS Publication No. SMA 00-3466, NHSDA Series H-12). Rockville, MD: Substance Abuse and Mental Health Services Administration.

Office of Applied Studies. 2001. *Summary of findings from the 2000 National Household Survey on Drug Abuse* (DHHS Publication No. SMA 00-3466, NHSDA Series H-12). Rockville, MD: Substance Abuse and Mental Health Services Administration.

Office of Applied Studies. 2003. *Summary of findings from the 2002 National Household Survey on Drug Abuse* (DHHS Publication No. SMA 00-3466, NHSDA Series H-12). Rockville, MD: Substance Abuse and Mental Health Services Administration.

Office of Applied Studies. 2004. *Summary of findings from the 2003 National Household Survey on Drug Abuse.* (DHHS Publication No. SMA 00-3466,

NHSDA Series H-12). Rockville, MD: Substance Abuse and Mental Health Services Administration.

Office of National Drug Control Policy. October 2003. What Americans need to know about marijuana: Important facts about our nation's most misunderstood illegal drug. *www.ncjrs.gov/ondcppubs/publications/pdf/mj_rev.pdf*

Pihl, R. O., and J. B. Peterson. 1995. Alcoholism: The role of different motivational systems. *Journal of Psychiatry and Neuroscience* 20 (5): 372–396.

Pitts, M., ed. 2003. *Institutional animal care and use committee guidebook*. Collingdale, PA: DIANE Publishing.

Understanding Animal Research. 2005. Animal research facts: Number of animals, types of animals, the research process, areas of research, and myth and fact. *www.rds-online.org.uk/pages/page.asp?i_ToolbarID=2&i_PageID=2*

Understanding Animal Research. 2005. Pain, distress and suffering. *www.rds-online.org.uk/pages/page.asp?i_ToolbarID=5&i_PageID=159*

Reich, W. T., ed. 1995. *Encyclopedia of bioethics*. New York: Macmillan Publishing Company.

Roberts, L. W., and A. R. Dyer. 2004. *Ethics in mental health care*. Washington, DC: American Psychiatric Publishing.

Roberts, T. G., G. P. Fournet, and E. Penland. 1995. A comparison of the attitudes toward alcohol and drug use and school support by grade level, gender, and ethnicity. *Journal of Alcohol and Drug Education* 40: 112–127.

Rowen, A. N. 1997. The benefits and ethics of animal research. *Scientific American* 276 (2): 79.

Schuckit, M., and J. Crabbe. Genetics of alcoholism susceptibility and protection. PowerPoint Presentation of Genetic Influences in Alcoholism.

Sperry, S.R., ed. 2000. For the greater good. Washington Association for Biomedical Research (Special Issue). *Seattle Post-Intelligencer.*

Twomey Fosnot, C., ed. 1996. *Constructivism: Theory, perspectives, and practice*. New York: Teachers College Press.

U.S. Census Bureau. 2002. Census 2000 Summary File 1 (SF 1) 100-percent data. *http://factfinder.census.gov*

U.S. Department of Agriculture. 1985. *Animal welfare act as amended* (7USC, 2131-2156).

U.S. Department of Justice National Drug Intelligence Center. Fast facts. *http://www.usdoj.gov/ndic/topics/ffacts.htm*

U.S. Drug Enforcement Administration (DEA). Drug descriptions. *www.usdoj.gov/dea/concern/concern.htm*

Vannin, E. 1999. Human experiments. *www.worldnewsstand.net/health/humanexperiments.htm*

Washington University School of Medicine. Neuroscience for kids: Modeling the nervous system. *http://faculty.washington.edu/chudler/chmodel.html*

The White House Office of National Drug Control Policy. 1995. *Drug control strategy*. Washington, DC: Author.

Wisconsin Association for Biomedical Research & Education (WABRE). Bioscience Wisconsin 2004 report. *www.wabre.org/2004bioscience/index.html*

Wu, T. C., et al. 1998. Pulmonary hazards of smoking marijuana as compared with tobacco. *New England Journal of Medicine* 318 (6): 347–351.

Yadid, G. 2005. Understanding through animal models. *CNS Spectrums* 10 (3): 181.

Yudofsky S., and R. E. Hales. 2010. *Essentials of neuropsychiatry and behavioral neurosciences.* Washington DC: American Psychiatric Press.

INDEX

INDEX

Axon terminals, 60, 65–66, 69, 70, 71, 77, 78, 219, 220, 251

B

Bar graphs, 39, 50, 214
Behavioral research, 8, 92, 94, 100, 251
Beneficence, ethical principle of, 7, 133, 135, 137, 146, 152, 153, 232, 237–238, 251
Bioethicist, 252
Bioethics, 252. *See also* Ethics
Biological research, 8, 92, 93, 94, 100, 252
Biological Sciences Curriculum Study (BSCS) 5E Instructional Model, 10, 12
Biomedical research, 144, 252
Blood pressure, 252
Blood work, 181, 184, 244
Board game, as assessment option, 204
Bookmark: Steps of the Scientific Method, 41, 54
Brain, 252
    cerebral cortex of, 252
    effects of drugs of abuse on, 8, 75–88 (*See also* Lesson 4)
    facts about, 57
    frontal lobe of, 64, 68, 218, 254
    limbic system of, 81, 255
    motor cortex of, 256
    occipital lobe of, 64, 68, 218, 256
    parietal lobe of, 64, 68, 218, 257
    reward system of, 80, 257
    structure and function of, 8, 55–73 (*See also* Lesson 3)
    temporal lobe of, 64, 68, 218, 258
Brain imaging technologies, 9, 180, 181, 184, 244, 253, 255, 257
Brainstem, 64, 68, 218, 252

C

Categories of neuroscience research, 8, 89, 92, 94, 100
Cats, inclusion in research, 171, 175
Cell, 252
Cell body (soma), 60, 65, 69, 70, 77, 78, 219, 252, 258

Cell culture, 173, 181, 184, 244
Centers for Disease Control and Prevention (CDC), 5, 151
Central nervous system, 252
Cerebellum, 64, 68, 218, 252
Cerebral cortex, 252
Cerebrum, 252
Characteristic, 252
Chemical message, 60, 194, 200, 223, 253
Chemical simulations, 182, 184, 244
Chronic condition, 253
Classify, 253
Cocaine, 53, 79, 82, 217, 253
Commercial, as assessment option, 204
Committee on Animal Research and Ethics (CARE), 159
Compare, 253
Compassion, ethical principle of, 7, 133, 135, 137, 146, 152, 153, 232, 237–238, 253
Computed tomography (CT); computer axial tomography (CAT), 253
Computer simulations, 9, 173, 180, 184, 244
Concept map, 253
Conclude, 253
Conclusion, 253
Consortium of National Arts Education Associations Standards, vii
    in Lesson 1, 24
Contrast, 253
Control group, 93, 253
Crack cocaine, 53, 217
Craving, 80, 253
Critical-thinking skills, 8, 75, 125, 127, 145, 157
Crossword puzzle, 62, 72–73
    answer key for, 221–222
Cruelty to Animals Act (England 1876), 143

D

Data, 253
    graphs of, 39–40, 45, 50–51
Decision-making skills, 3, 8, 23, 35, 40, 142, 157
Delta-9-tetrahydrocannabinol (THC), 255, 258
Dendrites, 60, 65, 69, 70, 77, 78, 219, 253
Disease, 253

INDEX

INDEX

INDEX

for students, 271–273
for teachers, 262–266
Review Board Committee, 144
Reward, 80, 257
Rubrics
for poster presentation, 205–207
for unit test essay question, 203

S

Sanitize, 258
Scientific American, 109
Scientific method, 3, 8, 37–38, 91, 258. *See also*
Lesson 2
bookmark for steps of, 41, 54
worksheet on, 47
answer key for, 211
worksheet on using, 48–49
answer key for, 212–213
Seizure, 258
Serotonin, 258
Skit about risks of drug abuse, 82
Soma, 60, 65, 69, 70, 77, 78, 219, 252, 258
Spinal column, 258
Spinal cord, 258
Sterile, 258
Steroids, 53, 82, 217, 258
Stimulants, 258
Student glossary, 27, 107, 251–260
Student project portfolio, 27
cover page for, 27, 31, 208
Student resources, 271–273
Synapses, 60, 65, 78, 219, 258
sending messages across, 66, 69, 71, 220
unnatural amount of neurotransmitters in,
79–80

T

Teacher resources, 262–266. *See also specific
lessons*
Teen profile, 8. *See also* Lesson 1
Temporal lobe of brain, 64, 68, 218, 258
Terra Nova Learning Systems (TNLS), vii
Thalamus, 258

This Is Your Brain unit. *See also* Lessons 1 to 10
assessments for, 15, 189–247
correlation with National Science Education
Standards, 11
description of, 3
development of, vii, 3
flexibility of, 14
format and timeline of, 13
goals and objectives of, 7
homework extensions for (*See* Homework
extensions)
material supply list for, 16–17
overview of lessons in, 8–9
parent letter explaining, 13, 28, 261
supporting materials for, 250–277
transparencies for (*See* Transparencies)
vocabulary terms introduced in, 18–19
worksheets for (*See* Worksheets)
Three Rs of animal inclusion in research, 7, 9,
159, 173, 175–176, 184, 244, 259
Tissue culture, 181, 184, 244
Tobacco use, 145, 146, 151, 238, 256, 259
Tolerance, 259
Transparencies
Animal Care and Use Committee Alerted
(Q), 147, 149, 160
Animal Research Reduced at Local Lab
(S), 175, 179
Can a Mouse Help Chris? (M), 94, 98, 105,
144
Can Research Help Chris? (K), 82, 84, 91
Chris Collapses in Gym! (A), 25, 29
Chris Released From Hospital (R), 164,
166, 177
Chris Tests Positive for Drugs! (C), 37, 43
Chris Treated at Emergency Room (B), 27,
30, 37
Doctor Questions Chris's Best Friend (N),
109, 115, 128
Functions of the Brain (G), 57, 58, 64
How a Neuron Sends a Message (H), 59,
62
How Have Drugs Affected Chris's Brain?
(J), 62, 67, 77
Key Ethics Terms (O), 129, 133